DOE/EIA-0554(2000)

Assumptions to the
Annual Energy Outlook 2000
(AEO2000)

With Projections to 2020

January 2000

www.eia.dce.gov

Contents

Introduction

This paper presents the major assumptions of the National Energy Modeling System (NEMS) used to generate the projections in the *Annual Energy Outlook 2000*[1] (*AEO2000*), including general features of the model structure, assumptions concerning energy markets, and the key input data and parameters that are most significant in formulating the model results. Detailed documentation of the modeling system is available in a series of documentation reports.[2] A synopsis of NEMS, the model components, and the interrelationships of the modules is presented in *The National Energy Modeling System: An Overview*.[3]

The National Energy Modeling System

The projections in the *AEO2000* were produced with the National Energy Modeling System. NEMS is developed and maintained by the Office of Integrated Analysis and Forecasting of the Energy Information Administration (EIA) to provide projections of domestic energy-economy markets in the midterm time period and perform policy analyses requested by decisionmakers and analysts in the U.S. Congress, the Department of Energy's Office of Policy, other DOE offices, and other government agencies.

The time horizon of NEMS is approximately 20 years, the midterm period in which the structure of the economy and the nature of energy markets are sufficiently understood that it is possible to represent considerable structural and regional detail. Because of the diverse nature of energy supply, demand, and conversion in the United States, NEMS supports regional modeling and analysis in order to represent the regional differences in energy markets, to provide policy impacts at the regional level, and to portray transportation flows. The level of regional detail for the end-use demand modules is the nine Census divisions. Other regional structures include production and consumption regions specific to oil, gas, and coal supply and distribution, the North American Electric Reliability Council regions and subregions for electricity, and aggregations of the Petroleum Administration for Defense Districts (PADD) for refineries. Only national results are presented in the *AEO2000,* with the regional and other detailed results available on the EIA CD-ROM and EIA Home Page. (http://www.eia.doe.gov/oiaf/aeo/index.html)

For each fuel and consuming sector, NEMS balances the energy supply and demand, accounting for the economic competition between the various energy fuels and sources. NEMS is organized and implemented as a modular system (Figure 1). The modules represent each of the fuel supply markets, conversion sectors, and end-use consumption sectors of the energy system. NEMS also includes macroeconomic and international modules. The primary flows of information among each of these modules are the delivered prices of energy to the end user and the quantities consumed by product, region, and sector. The delivered prices of fuel encompass all the activities necessary to produce, import, and transport fuels to the end user. The information flows also include other data such as economic activity, domestic production activity, and international petroleum supply availability.

The integrating module of NEMS controls the execution of each of the component modules. To facilitate modularity, the components do not pass information to each other directly but communicate through a central data storage location. This modular design provides the capability to execute modules individually, thus allowing decentralized development of the system and independent analysis and testing of individual modules. This modularity allows use of the methodology and level of detail most appropriate for each energy sector. NEMS solves by calling each supply, conversion, and end-use demand module in sequence until the delivered prices of energy and the quantities demanded have converged within tolerance, thus achieving an economic equilibrium of supply and demand in the consuming sectors. Solution is reached annually through the midterm horizon. Other variables are also evaluated for convergence such as petroleum product imports, crude oil imports, and several macroeconomic indicators.

Each NEMS component also represents the impact and cost of legislation and environmental regulations that affect that sector. NEMS reflects all current legislation and environmental regulations, such as the Clean Air Act Amendments of 1990 (CAAA90), and the costs of compliance with other regulations. NEMS also includes an analysis of the impacts of the provisions of the Climate Change Action Plan (CCAP), which are separately described under each module.

Figure 1. National Energy Modeling System

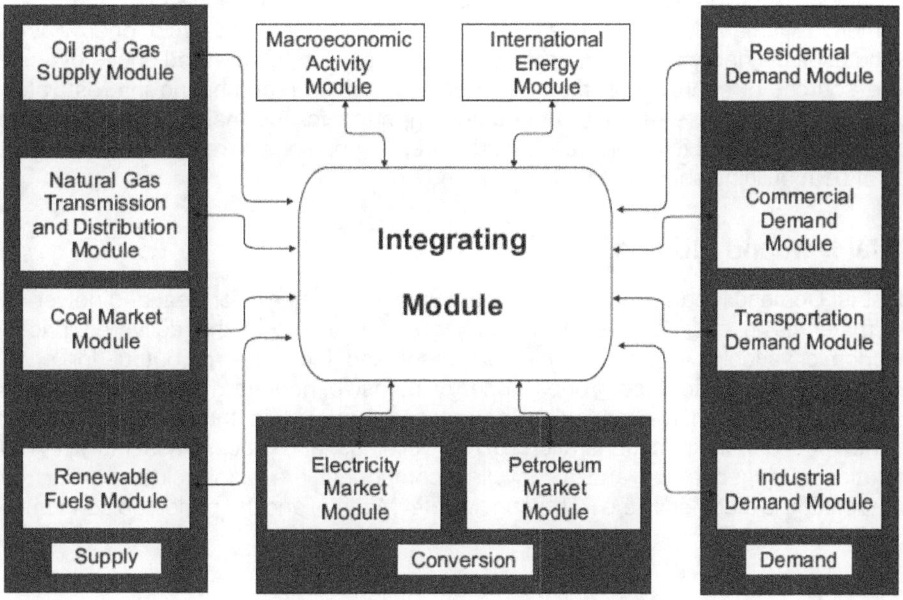

Component Modules

The component modules of NEMS represent the individual supply, demand, and conversion sectors of domestic energy markets and also include international and macroeconomic modules. In general, the modules interact through values representing the prices of energy delivered to the consuming sectors and the quantities of end-use energy consumption. This section provides brief summaries of each of the modules.

Macroeconomic Activity Module

The Macroeconomic Activity Module provides a set of essential macroeconomic drivers to the energy modules, and a macroeconomic feedback mechanism within NEMS. Key macroeconomic variables include gross domestic product (GDP), interest rates, disposable income, and employment. Industrial drivers are calculated for thirty-five industrial sectors. This module is a kernel regression representation of the Standard and Poor's DRI U.S. Macroeconomic Model of the U.S. Economy.

International Energy Module

The International Module represents the world oil markets, calculating the average world oil price and computing supply curves for five categories of imported crude oil for the Petroleum Market Module (PMM) of NEMS, in response to changes in U.S. import requirements. International petroleum product supply curves, including curves for oxygenates, are also calculated.

Household Expenditures Module

The Household Expenditures Module provides estimates of average household direct expenditures for energy used in the home and in private motor vehicle transportation. The forecasts of expenditures reflect the projections from NEMS for the residential and transportation sectors. The projected household energy expenditures incorporate the changes in residential energy prices and motor gasoline price determined in NEMS, as well as the changes in the efficiency of energy use for residential end-uses and in light-duty vehicle fuel efficiency. Average expenditures estimates are provided for households by income group and Census division.

Residential and Commercial Demand Modules

The Residential Demand Module forecasts consumption of residential sector energy by housing type and end use, subject to delivered energy prices, availability of renewable sources of energy, and housing starts. The Commercial Demand Module forecasts consumption of commercial sector energy by building types and nonbuilding uses of energy and by category of end use, subject to delivered prices of energy, availability of renewable sources of energy, and macroeconomic variables representing interest rates and floorspace construction. Both modules estimate the equipment stock for the major end-use services, incorporating assessments of advanced technologies, including representations of renewable energy technologies, and analyses of both building shell and appliance standards.

Industrial Demand Module

The Industrial Demand Module forecasts the consumption of energy for heat and power and for feedstocks and raw materials in each of sixteen industry groups subject to the delivered prices of energy and macroeconomic variables representing employment and the value of output for each industry. The industries are classified into three groups—energy intensive, nonenergy intensive, and nonmanufacturing. Of the eight energy-intensive industries, seven are modeled in the Industrial Demand Module with components for boiler/steam/cogeneration (BSC), buildings, and process/assembly (PA) use of energy. A representation of cogeneration and a recycling component are also included. The use of energy for petroleum refining is modeled in the Petroleum Market Module, and the projected consumption is included in the industrial totals.

Transportation Demand Module

The Transportation Demand Module forecasts consumption of transportation sector fuels, including petroleum products, electricity, methanol, ethanol, compressed natural gas, and hydrogen by transportation mode, vehicle vintage, and size class, subject to delivered prices of energy fuels and macroeconomic variables representing disposable personal income, GDP, population, interest rates, and the value of output for industries in the freight sector. Fleet vehicles are represented separately to allow analysis of the CAAA90 and other legislative proposals, and the module includes a component to explicitly assess the penetration of alternative-fuel vehicles.

Electricity Market Module

The Electricity Market Module (EMM) represents generation, transmission, and pricing of electricity, subject to delivered prices for coal, petroleum products, and natural gas, costs of generation by centralized renewables, macroeconomic variables for costs of capital and domestic investment, and electricity load shapes and demand. There are three primary submodules—capacity planning, fuel dispatching, finance and pricing. Nonutility generation and transmission and trade are represented in the planning and dispatching submodules. The levelized fuel cost of uranium fuel for nuclear generation is directly incorporated into the EMM. All CAAA90 compliance options are explicitly represented in the capacity expansion and dispatch decisions. Both new generating technologies and renewable technologies compete directly in these decisions.

Renewable Fuels Module

The Renewable Fuels Module (RFM) includes submodules that provide explicit representation of central station electricity supply for biomass (including wood, energy crops, and biomass co-firing), conventional hydroelectric, geothermal, municipal solid waste (including landfill gas), solar thermal, solar photovoltaics, and wind energy. The RFM contains natural resource supply estimates representing regional opportunities for renewable energy development.

Oil and Gas Supply Module

The Oil and Gas Supply Module represents domestic crude oil (including lease condensate), natural gas liquids, and natural gas supply within an integrated framework that captures the interrelationships among the various sources of supply—onshore, offshore, and Alaska—using both conventional and nonconventional techniques, including enhanced oil recovery and unconventional gas recovery from tight gas formations, shale, and coalbeds. This framework analyzes cash flow and profitability to compute investment and drilling in each of the supply sources, subject to the prices for crude oil and natural gas, the domestic recoverable resource base, and technology. Oil and gas production functions are computed at a level of twelve supply regions, including three offshore and three Alaskan regions. This module also represents foreign sources of natural gas, including pipeline imports and exports with Canada and Mexico, and liquefied natural gas imports and exports. Crude oil production quantities are input to the Petroleum Market Module in NEMS for conversion and blending into refined petroleum products. The supply curves for natural gas are input to the Natural Gas Transmission and Distribution Module for use in determining prices and quantities.

Natural Gas Transmission and Distribution Module

The Natural Gas Transmission and Distribution Module represents the transmission, distribution, and pricing of natural gas, subject to end-use demand for natural gas, the supply of domestic natural gas, and the availability of natural gas traded on the international market, on a seasonal basis. The module tracks the flow of natural gas in an aggregate, domestic pipeline network, connecting the domestic and foreign supply sources with twelve demand regions. This capability allows the analysis of impacts of interregional constraints in the interstate natural gas pipeline network and the identification of pipeline and storage capacity expansion requirements. Peak and off-peak periods are represented for natural gas transmission, and core and noncore markets are represented at the burner tip. The key components of pipeline and distributor tariffs are included in the pricing algorithms.

Petroleum Market Module

The Petroleum Market Module forecasts prices of petroleum products, crude oil and product import activity, and domestic refinery operations, including fuel consumption, subject to the demand for petroleum products, availability and price of imported petroleum, and domestic production of crude oil, natural gas liquids, and alcohol fuels. The module represents refining activities for three regions— Petroleum Administration for Defense District (PADD) 1, PADD 5, and an aggregate of PADDs 2, 3, and 4. The module uses the same crude oil types as the International Module. It explicitly models the requirements of CAAA90 and the costs of new automotive fuels, such as oxygenated and reformulated gasoline, and includes oxygenate production and blending for reformulated gasoline. *AEO2000* reflects the California ban on the gasoline blending component methyl tertiary butyl ether (MTBE) in 2003. Because the *AEO2000* reference case assumes current laws and regulations, it assumes that the Federal oxygen requirement for reformulated gasoline in Federal nonattainment areas will remain intact. Costs include capacity expansion for refinery processing units based on a 15-percent hurdle rate and a 15-percent return on investment. End-use prices are based on the marginal costs of production, plus markups representing product distribution costs, State and Federal taxes, and environmental costs.

Coal Market Module

The Coal Market Module represents mining, transportation, and pricing of coal, subject to the end-use demand for coal differentiated by physical characteristics, such as the heat and sulfur content. The coal supply curves include a response to fuel costs, labor productivity, and factor input costs. Twelve coal types are represented, differentiated by coal rank, sulfur content, and mining process. Production and distribution are computed for eleven supply and thirteen demand regions, using imputed coal transportation costs and trends in factor input costs. The Coal Market Module also forecasts the requirements for U.S. coal exports and imports. The international coal market component of the module computes trade in three types of coal for twenty import and sixteen export regions. Both the domestic and international coal markets are represented in a linear program.

Cases for the *Annual Energy Outlook 2000*

The *AEO2000* presents five cases which differ from each other due to fundamental assumptions concerning the domestic economy and world oil market conditions. Three alternative assumptions are specified for each of these two factors, with the reference case using the midlevel assumption for each.

- **Economic Growth** - In the reference case, productivity grows at an average annual rate of 1.3 percent from 1998 through 2020 and the labor force at 0.9 percent per year, yielding a growth in real GDP of 2.2 percent per year. In the high economic growth case, productivity and the labor force grow at 1.5 and 1.1 percent per year, respectively, resulting in GDP growth of 2.6 percent annually. The average annual growth in productivity, the labor force, and GDP is 1.0, 0.6, and 1.7 percent, respectively, in the low economic growth case.

- **World Oil Markets** - In the reference case, the average world oil price increases to $22.04 per barrel (in real 1998 dollars) in 2020. Reflecting uncertainty in world markets, the price in 2020 reaches $14.90 per barrel in the low oil price case and $28.04 per barrel in the high oil price case.

In addition to these five cases, additional cases presented in Table 1 explore the impacts of changing key assumptions in individual sectors.

Many of the side cases were designed to examine the impacts of varying key assumptions for individual modules or a subset of the NEMS modules, and thus the full market consequences, such as the consumption or price impacts, are not captured. In a fully integrated run, the impacts would tend to narrow the range of the differences from the reference case. For example, the best available technology side case in the residential demand assumed that all future equipment purchases are made from a selection of the most efficient technologies available in a particular year. In a fully integrated NEMS run, the lower resulting fuel consumption would have the effect of lowering slightly the market prices of those fuels with the concomitant impact of increasing economic growth, thus stimulating some additional consumption. As another example, the higher electricity demand side case results in higher electricity prices. If the end-use demand modules were executed in a full run, the demand for electricity would be reduced slightly as a result of the higher prices and resulting lower economic growth, thus moderating somewhat the input assumptions. The results of these cases should be considered the maximum range of the impacts that could occur with the assumptions defined for the case.

All projections are based on Federal, State, and local laws and regulations in effect on July 1, 1999, including the additional fuels taxes in the Omnibus Budget Reconciliation Act of 1993, the CAAA90, the Energy Policy Act of 1992, the Outer Continental Shelf Deep Water Royalty Relief Act of 1995, the Tax Payer Relief Act of 1997, and the Federal Highway Bill of 1998. Pending legislation and sections of existing legislation for which funds have not been appropriated are not reflected in these forecasts.

The projections include analysis of the provisions of the CCAP developed in 1993, which consists of forty-four actions to achieve carbon stabilization in the United States by 2000, relative to 1990. Thirteen of the actions not related to the combustion of energy fuels or to carbon dioxide and are not incorporated in the analysis. Since funding for many of the CCAP programs has been curtailed in budget negotiations, their full impact is not reflected in these projections. In addition, since some of the energy savings associated with CCAP programs are already in the baseline, the full projected impacts were reduced.

Emissions

Carbon emissions from energy use are dependent on the carbon content of the fuel and the fraction of the fuel consumed in combustion. The product of the carbon content at full combustion and the combustion fraction yields an adjusted carbon emission factor for each fuel. The emissions factors are expressed in millions of metric tons of carbon emitted per quadrillion Btu of energy use, or equivalently, in kilograms of carbon per million Btu. The adjusted emissions factors are multiplied by energy consumption to arrive at the carbon emissions projections.

For fuel uses of energy, the combustion fractions are assumed to be 0.99 for liquid fuels and 0.995 for gaseous fuels. The carbon in nonfuel use of energy, such as for asphalt and petrochemical feedstocks, is assumed to be sequestered in the product and not released to the atmosphere. For energy categories that are mixes of fuel and nonfuel uses, the combustion fractions are based on the proportion of fuel use. Any carbon emitted by renewable sources is considered balanced by the carbon sequestration that occurred in its creation. Therefore, following convention, net emissions of carbon from renewable sources is taken as zero, and no emission coefficient is reported. Renewable fuels include hydroelectric power, biomass, photovoltaic, geothermal, ethanol, and wind energy.

Table 2 presents the carbon coefficients at full combustion, the combustion fractions, and the adjusted carbon emission factors used for *AEO2000*.

Table 1.   Summary of *AEO2000* Cases

Case Name	Description	Integration mode
Reference	Baseline economic growth, world oil price, and technology assumptions	Fully Integrated
Low Economic Growth	Gross Domestic product grows at an average annual rate of 1.7 percent, compared to the reference case growth of 2.2 percent.	Fully Integrated
High Economic Growth	Gross domestic product grows at an average annual rate of 2.6 percent, compared to the reference case growth of 2.2 percent.	Fully Integrated
Low World Oil Price	World oil prices are $14.90 per barrel in 2020, compared to $22.04 per barrel in the reference case.	Fully Integrated
High World Oil Price	World oil prices are $28.04 per barrel in 2020, compared to $22.04 per barrel in the reference case.	Fully Integrated
Residential: 2000 Technology	Future equipment purchases based on equipment available in 2000. Building shell efficiencies fixed at 2000 levels.	Standalone
Residential: High Technology	Earlier availability, lower costs, and higher efficiencies assumed for more advanced equipment.	Standalone
Residential: Best Available Technology	Future equipment purchases and new building shells based on most efficient technologies available. Building shell efficiencies increase by 25 percent from1997 values by 2020.	Standalone
Commercial: 2000 Technology	Future equipment purchases based on equipment available in 2000. Building shell efficiencies fixed at 2000 levels.	Standalone
Commercial: High Technology	Earlier availability, lower costs, and higher efficiencies assumed for more advanced equipment.	Standalone
Commercial: Best Available Technology	Future equipment purchases based on most efficient technologies available. Building shell efficiencies increase 50 percent from reference values by 2020.	Standalone
Buildings: 10-Percent Standards	Assumes that near-term standards will be promulgated on time, and that future revisions will increase efficiency by 10 percent if technically feasible.	Standalone
Buildings 20-percent Standards	Assumes that near-term standards will be promulgated on time, and that future revisions will increase efficiency by 20 percent if technically feasible.	Standalone
Industrial: 2000 Technology	Efficiency of plant and equipment fixed at 2000 levels.	Standalone
Industrial: High Technology	Earlier availability, lower costs, and higher efficiencies assumed for more advanced equipment.	Standalone
Transportation: 2000 Technology	Efficiencies for new equipment in all modes of travel are fixed at 2000 levels.	Standalone
Transportation: High Technology	Reduced costs and improved efficiencies are assumed for advanced technologies.	Standalone
Consumption: 2000 Technology	Combination of the residential, commercial, industrial, and transportation 2000 technology cases and electricity low fossil technology case.	Fully Integrated

Table 1. Summary of the *AEO2000* Cases (continued)

Case Name	Description	Integration mode
Consumption: High Technology	Combination of the residential, commercial, industrial, and transportation high technology cases and electricity high fossil technology case.	Fully Integrated
Electricity: Low Nuclear	Relative to the reference case, higher capital investments and operating costs are assumed to be required when plants are evaluated for retirement.	Partially Integrated
Electricity: High Nuclear	No capital investments are assumed to be required through the initial 40-year plant lifetime, and investment and operating costs are lower than in the reference case.	Partially Integrated
Electricity: High Demand	Electricity demand increases at an annual rate of 2.0 percent, compared to 1.4 percent in the reference case.	Partially Integrated
Electricity: Low Fossil Technology	New fossil generating technologies are assumed not to improve over time from 1999.	Fully Integrated
Electricity: High Fossil Technology	Costs and efficiencies for advanced fossil-fired generating technologies are assumed to improve from reference case values.	Fully Integrated
Electricity: Competitive Pricing With Reference Gas Prices	Competitive pricing is phased in over 10 years in all regions of the country.	Fully Integrated
Electricity: Competitive Pricing With Higher Gas Prices	Competitive pricing is phased in over 10 years in all regions of the country. Cost, finding rate, and success rate parameters for natural gas adjusted for slower improvement.	Fully Integrated
Electricity: Competitive Pricing With Lower Gas Prices	Competitive pricing is phased in over 10 years in all regions of the Country. Cost, finding rate, and success rate parameters for natural gas adjusted for more rapid improvement.	Fully Integrated
Electricity: RPS with Cap and Sunset	Nonhydroelectric renewable generation is required to increase to 7.5 percent of total electricity sales for the period 2010-2015, subject to a 1.5 cent per kilowatthour limit on the price of of renewable credits. The RPS requirement sunsets in 2015.	Fully Integrated
Electricity: RPS with Cap, No Sunset	Nonhydroelectric renewable generation is required to increase to 7.5 percent of total electricity sales in 2010 and all years thereafter, subject to a 1.5 cent per kilowatthour limit on the price of renewable credits.	Fully Integrated
Electricity: RPS, No Cap, No Sunset	Nonhydroelectric renewable generation is required to increase to 7.5 percent of total electricity sales in 2010 and all years thereafter.	Fully Integrated
Renewables: High Renewables	Lower costs and higher efficiencies are assumed for new renewable generating technologies.	Fully Integrated
Oil and Gas: Slow Technology	Cost, finding rate, and success rate parameters adjusted for slower improvement.	Fully Integrated
Oil and Gas: Rapid Technology	Cost, finding rate, and success rate parameters adjusted for more rapid improvement.	Fully Integrated

Table 1. Summary of the *AEO2000* Cases (continued)

Case Name	Description	Integration mode
Oil and Gas: Gasoline Sulfur Reduction	The sulfur content of all gasoline in the United States is reduced to a 30 ppm annual average standard starting in 2004. Reformulated gasoline meets the 30 ppm requirement in 2004. Conventional gasoline meets an 80 ppm specification in 2004 but meets the 30 ppm limit by 2007.	Standalone
Oil and Gas: BRP/MTBE Reduction	MTBE blended with gasoline is reduced to 3 percent of all gasoline by 2003. The Federal requirement for 2.0 percent oxygen in reformulated gasoline is waived.	Standalone
Coal: Low Mining Cost	Productivity increases at an annual rate of 3.6 percent, compared to the reference case growth of 2.3 percent. Real wages decrease by 0.5 percent annually, compared to constant real wages in the reference case.	Partially Integrated
Coal: High Mining Cost	Productivity increases at an annual rate of 0.9 percent, compared to the reference case growth of 2.3 percent. Real wages increase by 0.5 percent annually, compared to constant real wages in the reference case.	Partially Integrated

Table 2. Carbon Emission Factors (Kilograms-carbon per million Btu)

Fuel Type	Carbon Coefficient at Full Combustion	Combustion Fraction	Adjusted Emissions Factor
Petroleum			
Motor Gasoline	19.33	0.990	19.14
Liquefied Petroleum Gas			
Used as Fuel	17.20	0.995	17.11
Used as Feedstock	16.87	0.200	3.37
Jet Fuel	19.33	0.990	19.14
Distillate Fuel	19.95	0.990	19.75
Residual Fuel	21.49	0.990	21.28
Asphalt and Road Oil	20.62	0.000	0.00
Lubricants	20.24	0.600	12.14
Petrochemical Feedstocks	19.37	0.200	3.87
Kerosene	19.72	0.990	19.52
Petroleum Coke	27.85	0.500	13.93
Petroleum Still Gas	17.51	0.995	17.42
Other Industrial	20.31	0.990	20.11
Coal			
Residential and Commercial	25.92	0.990	25.66
Metallurgical	25.55	0.990	25.29
Industrial Other	25.61	0.990	25.39
Electric Utility[1]	25.74	0.990	24.486
Natural Gas			
Used as Fuel	14.47	0.995	14.40
Used as Feedstocks	14.47	0.774	11.20

[1]Emission factors for coal used for electricity generation are specified by coal supply region and types of coal, so the average carbon contents for coal varies throughout the forecast. The 1998 average is 24.486.

Source: Energy Information Administration, *Emissions of Greenhouse Gases in the United States 1998*, DOE/EIA-0573(98), (Washington, DC, October 1999).

Notes and Sources

[1] Energy Information Administration, *Annual Energy Outlook 2000* (AEO2000), DOE/EIA-0383(2000), (Washington, DC, December 1999).

[2] NEMS documentation reports are available on the EIA CD-ROM and the EIA Homepage (http://www.eia.doe.gov/bookshelf.html). For ordering information on the CD-ROM, contact STAT-USA's toll free order number: 1-800-STAT-USA or by calling (202) 482-1986.

[3] Energy Information Administration, *The National Energy Modeling System: An Overview* 1998, DOE/EIA-0581(98), (Washington, DC, February 1998).

Macroeconomic Activity Module

The Macroeconomic Activity Module (MAM) represents the interaction between the U.S. economy as a whole and energy markets. The rate of growth of the economy, measured by the growth in gross domestic product (GDP) is a key determinant of the growth in demand for energy. Associated economic factors, such as interest rates and disposable income, strongly influence various elements of the supply and demand for energy. At the same time, reactions to energy markets by the aggregate economy, such as a slowdown in economic growth resulting from increasing energy prices, are also reflected in this module. A detailed description of the MAM is provided in the EIA publication, *Model Documentation Report: Macroeconomic Activity Module (MAM) of the National Energy Modeling System*, DOE/EIA-M065, (Washington, DC, February 1994), plus *Macroeconomic Activity Module (MAM): Kernel Regression Documentation of the National Energy Modeling System 1999*, DOE/EIA-M065(99), Washington, DC, 1999).

Key Assumptions

The output of the Nation's economy, measured by GDP, is expected to increase by 2.2 percent between 1998 and 2020 in the reference case. The growth in GDP can be decomposed into two key factors: the growth rate of the labor force and the rate of productivity change associated with the labor force. As Table 3 indicates, the rate of growth of GDP is slower in the latter half of the forecast period due to a slowdown in the expansion of the labor force. The growth of the labor force depends upon the forecasted population growth and the labor force participation rate. The Census Bureau's middle series population projection is used as a basis for the *AEO2000*. Total population is expected to grow annually by 0.8 percent between 1998 and 2020, with a higher rate of growth pre-2000 and a slower rate of growth post-2000. Over the forecast period, the labor force participation rate is expected to peak in 2007 and then decline as "baby boom" cohorts begin to retire. Combining population projections with labor force participation rates gives an increase in labor force earlier in the forecast horizon and then post-2000, the economy experiences slower growth as demographic trends affect future economic growth.

Table 3. Growth in Gross Domestic Product, Labor Force, and Productivity
(Percent per Year)

Assumptions	1998-2000	2000-2005	2005-2010	2010-2015	2015-2020	1998-2020
GDP (Billion Chain-Weighted $1992)						
High Growth	3.4	3.2	2.4	2.3	2.3	2.6
Reference	2.9	2.5	2.1	2.1	1.8	2.2
Low Growth	1.9	2.3	1.6	1.5	1.2	1.7
Labor Force						
High Growth	1.6	1.4	1.2	0.8	0.8	1.1
Reference	1.4	1.1	1.0	0.7	0.5	0.9
Low Growth	1.0	1.0	0.7	0.4	0.3	0.6
Productivity						
High Growth	1.8	1.8	1.2	1.5	1.5	1.5
Reference	1.5	1.4	1.1	1.4	1.3	1.3
Low Growth	0.9	1.3	0.9	1.1	0.9	1.0

Source: Energy Information Administration, *AEO2000* National Energy Modeling System runs: aeo2k.d100199a; lmac2k.d100199a; and hmac2k.d100199a.

The productivity of labor is the second major reason for economic growth and reflects the positive effects of a growing capital stock of the economy as well as technological change occurring over time. A key to achieving the reference case's long-run 2.2 percent growth is an anticipated recovery in productivity growth. Productivity growth slowed in the 1970's, compared to the growth experienced post-World War II. There is no consensus about why productivity growth declined so much after 1973. However, between 1980 and 1990, business investment's share of GDP declined at the same time that both the Federal budget deficit and the trade deficit increased. Since 1991, the economic recovery has been led by strong gains in business investment as a result of lower interest rates. Productivity has shown recent strong gains as economic output has increased more rapidly than employment gains.

In the reference case, productivity growth remains relatively constant throughout the forecast period. Business fixed investment rises as a share of GDP. The resulting growth in the capital stock and the technology base of that capital stock helps to sustain productivity growth exceeding 1 percent. This growth in productivity offsets some of the decline in the labor force growth, but the economy continues to slow down over time.

To reflect the uncertainty in forecasts of economic growth, the *AEO2000* forecasts use high and low economic growth cases along with the reference case to project the possible energy markets. All three economic growth cases are based on forecasts prepared by Standard and Poor's DRI.[4] The DRI forecasts used in *AEO2000* are the August 1999 Trend Growth scenario along with the February 1999 Optimistic and Pessimistic growth projections.

The high economic growth case incorporates higher population, labor force and productivity growth rates than the reference case. Due to the higher productivity gains, inflation and interest rates are lower compared to the reference case. Investment, disposable income, and industrial production are increased. Economic output is projected to increase by 2.6 percent between 1998 and 2020. The low economic growth case assumes lower population, labor force, and productivity gains, with resulting higher prices and interest rates and lower industrial output growth. In the low economic growth case, economic output is expected to increase by 1.7 percent over the forecast horizon.

The regional disaggregation of the economic variables uses regional shares based on a regional model solution. These shares change over time, but do not change as energy prices change from the projected reference price path.

---

[4] The underlying macroeconomic growth cases use Standard and Poor's DRI August 1999 T250899 and February TO250299 and TP250299.

International Energy Module

The International Energy Module determines changes in the world oil price and the supply prices of crude oils and petroleum products for import to the United States in response to changes in U.S. import requirements. A market clearing method is used to determine the price at which worldwide demand for oil is equal to the worldwide supply. The module determines new values for oil production and demand for regions outside the United States, along with a new world oil price that balances supply and demand in the international oil market. A detailed description of the International Energy Module is provided in the EIA publication, *Model Documentation Report: The International Energy Module of the National Energy Modeling System*, DOE/EIA-M071(99), (Washington, DC, February 1999).

Key Assumptions

The level of oil production by countries in the OPEC is a key factor influencing the world oil price projections incorporated into *AEO2000*. Non-OPEC production, worldwide regional economic growth rates and the associated regional demand for oil are additional factors affecting the world oil price.

OPEC oil production is assumed to increase throughout the forecast, making OPEC the primary source, satisfying the worldwide increase in oil consumption expected over the forecast period (Figure 2). OPEC is assumed to be the source of additional production because its member nations hold a major portion of the world's total reserves—exceeding 800 billion barrels, over 77 percent of the world's estimated total, at the end of 1998.[5] For the *AEO2000* forecasts, three different OPEC production paths are the principal assumptions leading to the three world oil price path cases examined: the low oil price case, reference case, and high oil price case. The values assumed for OPEC production for the three world oil price cases are given in Figure 2. Iraq is assumed to continue selling oil only at United Nations Security Council sanction-allowed volumes until at least 2002. Once sanctions are lifted, Iraq will increase production levels to over 4 million barrels per day within 2 years. Within a decade of sanctions being lifted, Iraq is expected to increase production capacity to more than 6 million barrels per day with likely investment help from foreign sources. Non-OPEC oil production is expected to follow a gradually rising path—with an increase of more than 1.0 percent per year over the forecast period—as advances in both exploration and extraction technologies result in this upward trend (Figure 3). One fixed path for non-OPEC oil production is initially

Figure 2. OPEC Oil Production, 1970-2020
(Milliion Barrels per Day)

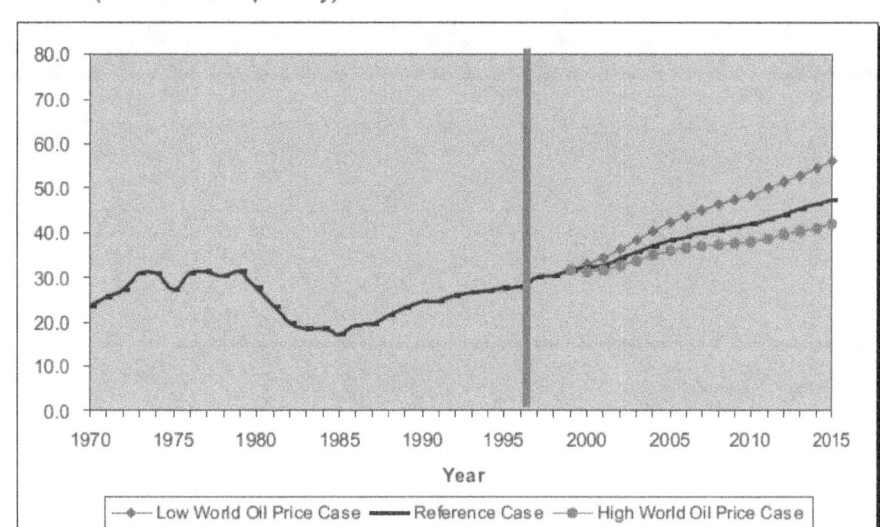

OPEC = Organization of Petroleum Exporting Countries.

Sources: Energy Information Administration. AEO2000 National Energy Modeling System runs lwop2k.d100199a, aeo2k.d100199a, and hwop2k.d100199a.

input for all three world oil price case projections. Non-OPEC production depends upon world oil prices, so the final forecast solutions of the levels of non-OPEC production for the three oil price cases diverge from the initial assumptions. Production is higher in the high oil price case since more marginal wells are profitable at the higher prices. Likewise, lower world oil prices are associated with lower production levels. The final non-OPEC production paths for the three oil price cases are shown in Figure 3.

Figure 3. Non-OPEC Oil Production, 1970-2020
(Million Barrels per Day)

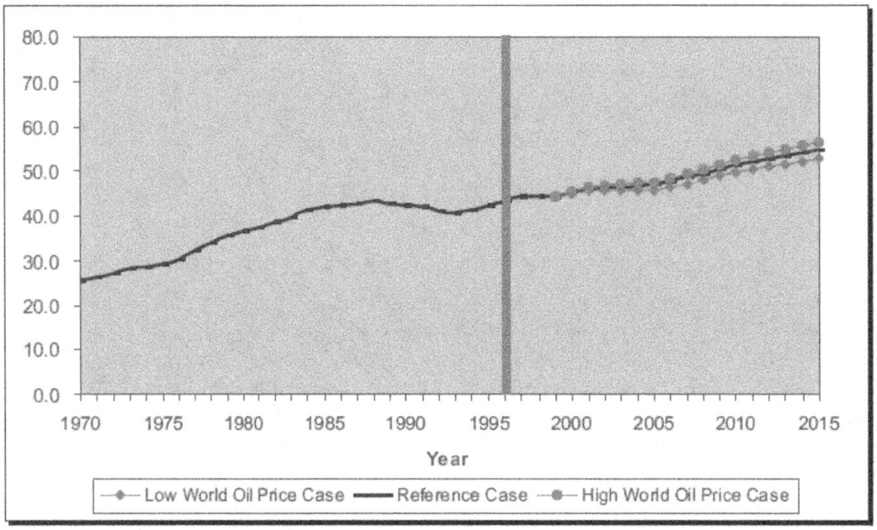

OPEC = Organization of Petroleum Exporting Countries.

Sources: Energy Information Administration. *AEO2000* National Energy Modeling System runs lwop2k.d100199a, aeo2k.d100199a, and hwop2k.d100199a.

The non-U.S. oil production forecasts in the AEO2000 begin with country-level assumptions regarding proved oil reserves. These reserve estimates are shown in Table 4 and are compiled by PennWell Publishing Company's Oil and Gas Journal.

Table 4. Worldwide Oil Reserves as of January 1, 1999
(Billion Barrels)

Region	Proved Oil Reserves
Western Hemisphere	164.8
Western Europe	18.7
Asia-Pacific	43.0
Eastern Europe and F.S.U.	59.1
Middle East	673.6
Africa	75.4
Total World	1,034.7
Total OPEC	800.5

Source: PennWell Publishing Co., International Petroleum Encyclopedia, (Tulsa, OK, 1999).

The assumed growth rates for GDP for various regions in the world are shown in Table 5. This set of growth rates for GDP was assumed for all three price cases. The GDP growth rate assumptions are from Standard & Poor's DRI third quarter 1999 World Economic Outlook.

The values for growth in oil demand calculated in the International Energy Module, which depend upon the oil price levels as well as the GDP growth rates, are shown in Table 6 for the three oil price cases by regions

Table 5. Average Annual Regional Gross Domestic Product Growth Rates, 1998-2020
(Percent per Year)

Region	Gross Domestic Product
Organization for Economic Cooperation and Development	1.9
Other Developing Countries	3.6
Eurasia	5.1
China	6.3
Former Soviet Union	4.3
Eastern Europe	4.4
Total World	2.8

Source: Standard & Poor's DRI, World Economic Outlook, Volume 1, (Lexington, MA, Third Quarter 1999).

Table 6. Average Annual Regional Growth Rates for Oil Demand, 1998-2020
(Percent per Year)

Region	Low Price	Reference	High Price
Organization for Economic Cooperation and Development	1.5	1.1	0.9
Organization of Petroleum Exporting Countries	2.3	2.3	2.3
Other Developing Countries	3.6	3.3	3.1
Eurasia	2.8	2.5	2.3
China	4.2	3.8	3.5
Former Soviet Union	1.8	1.5	1.4
Eastern Europe	1.0	0.9	0.8
Total World	2.2	1.9	1.7

Source: Energy Information Administration, *AEO2000* National Energy Modeling System runs: lwop2k.d100199a; aeo2k.d100199a; and hwop2k.d100199a.

Petroleum product imports are represented in the projections through a series of curves that present the quantity of each product that the world market is willing to supply to U.S. markets for each of the five Petroleum Administration for Defense Districts (PADDs). Curves are provided for ten products: traditional gasoline (including aviation), reformulated gasoline, No. 2 heating oil, low-sulfur distillate oil, high- and low-sulfur residual oil, jet fuel (including naptha jet), liquefied petroleum gas, petrochemical feedstocks, and other. The curves are calculated using the World Oil Refining Logistics Demand (WORLD) Model.[6] The WORLD model uses as inputs worldwide demand for crude oil and petroleum products for world oil prices that are close to the oil prices assumed for *AEO2000*, as well as values for worldwide petroleum production that are consistent with such prices. The refinery technology incorporated in the model is updated using the most recently available Oil & Gas Journal Database.[7]

Notes and Sources

[5]   PennWell Publishing Co., International Petroleum Encyclopedia, (Tulsa, OK, 1999).

[6]   EIA, *EIA Model Documentation: World Oil Refining Logistics Demand Model, "WORLD" Reference Manual,* DOE/EIA-M058, (Washington, DC, March 1994).

[7]   Oil & Gas Journal, *World Wide Refinery Survey*, (data as of January 1, 1999).

Household Expenditures Module

The Household Expenditures Module (HEM) constructs household energy expenditure profiles using historical survey data on household income, population and demographic characteristics, and consumption and expenditures for fuels for various end-uses. These data are combined with NEMS forecasts of household disposable income, fuel consumption, and fuel expenditures by end-use and household type. The HEM disaggregation algorithm uses these combined results to forecast household fuel consumption and expenditures by income quintile and Census Division.

Key Assumptions

The historical input data used to develop the HEM version for the *AEO2000* consists of recent household survey responses, aggregated to the desired level of detail. Two surveys performed by the Energy Information Administration are included in the *AEO2000* HEM database, and together these input data are used to develop a set of baseline household consumption profiles for the direct fuel expenditure analysis. These surveys are the 1997 Residential Energy Consumption Survey (RECS) and the 1991 Residential Transportation Energy Consumption Survey (RTECS).

HEM uses the consumption forecast by NEMS for the residential and transportation sectors as inputs to the disaggregation algorithm that results in the direct fuel expenditure analysis. Household end-use and personal transportation service consumption are obtained by HEM from the NEMS Residential and Transportation Demand Modules. Household disposable income is adjusted with forecasts of total disposable income from the NEMS Macroeconomic Activity Module.

The fundamental assumptions underlying HEM's processing of the historical and NEMS forecast data to obtain its results are:

- Individual households are assumed not to migrate between income quintiles throughout the analysis period.

- All households within a household segment are assumed to consume the average quantity of fuel for that segment. Distributions about, or deviations from, the average are not explicitly modeled.

- The change in average household consumption between forecast year y and survey base year y_0 is captured from the NEMS run at the finest available level of detail, and the same proportional change is assumed to occur in each HEM subsegment of the analysis.

Application of the HEM algorithm produces a direct household fuel expenditure forecast at the finest level of disaggregation; namely, by fuel, end-use service, housing type and vintage, ethnicity, disposable income quintile, Census Division, and year. Results obtained are summed across end-uses to yield total direct fuel expenditures as a function of disposable income for each household segment. The consolidation of these high-resolution results into national average household expenditure results requires a weighted averaging in order to obtain the desired aggregations. The weighing scheme used requires the proportions of households of each type and vintage headed by householders of each ethnicity and income quintile. The survey data provides these historical subsegment proportions, and for the *AEO2000* they are assumed to remain constant throughout the forecast period.

Residential Demand Module

The NEMS Residential Demand Module forecasts future residential sector energy requirements based on projections of the number of households and the stock, efficiency, and intensity of use of energy-consuming equipment. The Residential Demand Module projections begin with a base year estimates of the housing stock, the types and numbers of energy-consuming appliances servicing the stock, and the "unit energy consumption" by appliance (or UEC—in million Btu per household per year). The projection process adds new housing units to the stock, determines the equipment installed in new units, retires existing housing units, and retires and replaces appliances. The primary exogenous drivers for the module are housing starts by type (single-family, multifamily and mobile homes) and Census Division and prices for each energy source for each of the nine Census Divisions. The Residential Demand Module also requires projections of available equipment over the forecast horizon. Over time, equipment efficiency tends to increase because of general technological advances and also because of Federal and/or state efficiency standards. As energy prices and available equipment changes over the forecast horizon, the module includes projected changes to the type and efficiency of equipment purchased as well as projected changes in the usage intensity of the equipment stock.

The end-use services for which equipment stocks are modeled include space conditioning (heating and cooling), water heating, refrigeration, freezers, dishwashers, clothes washers, lighting, furnace fans, cooking, and clothes drying. In addition to the major equipment-driven end-uses, the average energy consumption per household is projected for secondary heating, color televisions, personal computers, and other electric and nonelectric appliances. The module's output includes number of households, equipment stock, average equipment efficiencies, and energy consumed by service, fuel, and geographic location. The fuels represented are distillate fuel oil, liquefied petroleum gas, natural gas, kerosene, electricity, wood, geothermal, coal, and solar energy.

One of the implicit assumptions embodied in the Residential Demand Module is that through 2020, there will be no radical changes in technology or consumer behavior. No new regulations of efficiency beyond those currently embodied in law or new government programs fostering efficiency improvements are assumed. Technologies which have not gained widespread acceptance today, will not achieve significant penetration by 2020. Currently available technologies will evolve in both efficiency and cost. In general, for the same real cost, future technologies will be less expensive than those available today. When choosing new or replacement technologies, consumers will behave similarly to the way they now behave. The intensity of end-uses will change moderately in response to price changes. Electric end uses will continue to expand, but at a decreasing rate[8]

Key Assumptions

Housing Stock Submodule

A very important determinant of future energy consumption is the projected number of households. Base year estimates for 1997 are derived from the Energy Information Administration's (EIA) *Residential Energy Consumption Survey* (RECS) (Table 7). The forecast for occupied housing units is done separately for each Census Division. It is based on the combination of the previous year's surviving stock with projected housing starts provided by the NEMS Macroeconomic Activity Module. The housing stock submodule assumes a constant survival rate (the percentage of households which are present in the current forecast year, which were also present in the preceding year) for each type of housing unit; 99.6 percent for single-family units, 99.6 percent for multifamily units, and 96.5 percent for mobile home units. Projected fuel consumption is dependent not only on the projected number of housing units, but also on the type and geographic distribution of the houses. The intensity of space heating energy use varies greatly across the various climate zones in the United States. Also, fuel prevalence varies across the country—oil (distillate) is more frequently used as a heating fuel in the New England and Middle Atlantic Census Divisions than in the rest of the country, while natural gas dominates in the Midwest. An example of differences by housing type is the more prevalent use of liquefied petroleum gas in mobile homes relative to other housing types.

Table 7. 1997 Households

Region	Single-family Units	Multi-family Units	Mobile Home Units	Total Units
New England	3,759,905	1,434,960	114,801	5,309,666
Mid Atlantic	9,990,266	4,063,826	370,168	14,424,260
East North Central	12,541,488	3,616,338	748,928	16,906,754
West North Central	5,905,676	893,549	353,749	7,152,974
South Atlantic	13,638,587	3,566,115	1,488,834	18,693,536
East South Central	4,785,180	769,795	788,963	6,343,938
West South Central	8,231,512	1,899,383	708,128	10,839,023
Mountain	4,476,532	1,039,756	663,026	6,179,314
Pacific	10,406,761	4,144,606	1,080,339	15,631,706
United States	73,735,907	21,428,328	6,316,936	101,481,171

Source: Energy Information Administration, *A Look at Residential Energy Consumption in 1997*, DOE/EIA-314(97), (Washington, DC, November 1999).

Technology Choice Submodule

The key inputs for the Technology Choice Submodule are fuel prices by Census Division and characteristics of available equipment (installed cost, maintenance cost, efficiency and equipment life). Fuel prices are determined by an equilibrium process which considers energy supplies and demands and are passed to this submodule from the integrating module of NEMS. Energy price, combined with equipment UEC (which is a function of efficiency), determines the operating costs of equipment. Equipment characteristics are exogenous to the model and are modified to reflect both Federal standards and anticipated changes in the market place. Table 8 lists capital cost and efficiency for selected residential appliances for the years 1998 and 2005.

Table 8. Installed Cost and Efficiency Ratings of Selected Equipment

Equipment Type	Relative Performance[1]	1998 Installed Cost ($1998)[2]	Efficiency[3]	2015 Installed Cost ($1998)[2]	Efficiency[3]	Approximate Discount Rate
Electric Heat Pump	Minimum	$4,100	10.0	$4,100	10.0	38%
	Best	$5,555	17.7	$5,200	18.0	
Natural Gas Furnace	Minimum	$1,300	0.78	$1,300	0.78	15%
	Best	$2,700	0.96	$1,600	0.96	
Room Air Conditioner	Minimum	$450	8.7	$450	9.7	140%
	Best	$760	11.7	$760	12.0	
Central Air Conditioner	Minimum	$2,500	10.0	$2,500	10.0	36%
	Best	$3,600	18.0	$3,200	18.0	
Refrigerator (18 cubic ft)	Minimum	$530	690	$530	478	19%
	Best	$850	518	$700	400	
Electric Water Heater	Minimum	$350	0.86	$350	0.86	83%
	Best	$1,025	2.60	$800	2.20	
Solar Water Heater	N/A	$2,600	2.0	$2,600	2.0	83%

[1]Minimum performance refers to the lowest efficiency equipment available. Best refers to the highest efficiency equipment available.

[2]Installed costs are given in 1998 dollars.

[3]Efficiency measurements vary by equipment type. Electric heat pumps and central air conditioners are rated for cooling performance using the Seasonal Energy Efficiency Ratio (SEER); natural gas furnaces are based on Annual Fuel Utilization Efficiency; room air conditioners are based on Energy Efficiency Ratio (EER); refrigerators are based on kilowatt-hours per year; and water heaters are based on Energy Factor (delivered Btu divided by input Btu).

Source: Arthur D. Little, *EIA Technology Forecast Updates*, Reference Number 37125, September 1998.

The Residential Demand Module projects equipment purchases based on a nested choice methodology. The first stage of the choice methodology determines the fuel and technology to be used, the second stage determines the efficiency of the selected equipment type. For new construction, home heating fuel and technology choices are determined based on life-cycle costs assuming a 20 percent discount rate. The equipment choices for cooling, water heating, and cooking are linked to the space heating choice for new construction. Technology and fuel choice for replacement equipment uses a nested methodology similar to that for new construction, but includes (in addition to the capital and installation costs of the equipment), explicit costs for technology switching (e.g., costs for installing gas lines if switching from electricity or oil to gas, or costs for retrofitting air ducts if switching from electric resistance heat to central heating types). Also, for replacements, there is no linking of fuel choice for water heating and cooking as is done for new construction. Technology switching upon replacement is allowed for space heating, air conditioning, water heating, cooking and clothes drying.

Once the fuel and technology choice for a particular end use is determined, the second stage of the choice methodology determines efficiency. In any given year, there are several available prototypes of varying efficiency (minimum standard, medium low, medium high and highest efficiency). Efficiency choice is based on a functional form and coefficients which give greater or lesser importance to the installed capital cost (first cost) versus the operating cost. Generally, within a technology class, the higher the first cost, the lower the operating cost.

The parameters for the second stage efficiency choice are calibrated to the most recently available shipment data for the major residential appliances. Shipment efficiency data are obtained from industry associations which monitor shipments such as the Association of Home Appliance Manufacturers. Because of this calibration procedure, the model allows the relative importance of first cost versus operating cost to vary by general technology and fuel type (e.g., natural gas furnace, electric heat pump, electric central air conditioner, etc.). Once the model is calibrated, it is possible to calculate (approximately) the apparent discount rates based on the relative weight given to the operating cost savings versus the weight given to the higher cost of more efficient equipment. Discount rates in excess of 30 percent are common in the Residential Demand Module. The prevalence of such high apparent discount rates by consumers has led to the notion of the "efficiency gap"— that is, there are many investments that could be made that provide rates of return in excess of residential borrowing rates (15 to 20 percent for example). There are several studies which document instances of apparent high discount rates.[9] The efficiency gap literature has been drawn on as the basis for efficiency standards and Federally-Sponsored voluntary programs under the Climate Change Action Plan (CCAP) (see on page 25). Once equipment efficiencies for a technology and fuel are determined, the installed efficiency for its entire stock is calculated.

Appliance Stock Submodule

The Appliance Stock Submodule is an accounting framework which tracks the quantity and average efficiency of equipment by end use, technology, and fuel. It separately tracks equipment requirements for new construction and existing housing units. For existing units, this module calculates equipment which survives from previous years, allows certain end uses to further penetrate into the existing housing stock and calculates the total number of units required for replacement and further penetration. Air conditioning and clothes drying are the two end uses not considered to be "fully penetrated."

Once a piece of equipment enters into the stock, an accounting of its remaining life is begun. It is assumed that all appliances survive a minimum number of years after installation. A fraction of appliances are removed from the stock once they have survived for the minimum number of years. Between the minimum and maximum life expectancy, all appliances retire based on a linear decay function. For example, if an appliance has a minimum life of 5 years and a maximum life of 15 years, one tenth of the units (1 divided by 15 minus 5) are retired in each of years 6 through 15. It is further assumed that, when a house is retired from the stock, all of the equipment contained in that house retires as well; i.e., there is no secondhand market for this equipment. The assumptions concerning equipment lives are given in Table 9.

Fuel Consumption Submodule

Energy consumption is calculated by multiplying the vintage equipment stocks by their respective UECs. The UECs include adjustments for the average efficiency of the stock vintages, short term price elasticity of demand and "rebound" effects on usage (see discussion on page 24), the size of new construction relative to the existing stock, people per household and shell efficiency and weather effects (space heating andcooling). The various levels of aggregated consumption (consumption by fuel, by service, etc.) are derived from these detailed equipment-specific calculations.

Table 9. Minimum and Maximum Life Expectancies of Equipment

Equipment	Minimum Life	Maximum Life
Heat Pumps	8	16
Central Forced-Air Furnaces	18	29
Hydronic Space Heaters	20	30
Room Air Conditioners	12	19
Central Air Conditioners	8	16
Water Heaters	12	19
Cooking Stoves	16	21
Clothes Dryers	6	30
Refrigerators	7	26
Freezers	11	31

Source: Lawrence Berkeley Laboratory, *Baseline Data for the Residential Sector and Development of a Residential Forecasting Database*, May 1994, and analysis of RECS 1997 data.

Equipment Efficiency

The average energy consumption of a particular technology is initially based on estimates derived from RECS 1997. Appliance efficiency is either derived from a long history of shipment data (e.g., the efficiency of conventional air-source heat pumps) or assumed based on engineering information concerning typical installed equipment (e.g., the efficiency of ground-source heat pumps). When the average efficiency is computed from shipment data, shipments going back as far as 20 to 30 years are combined with assumptions concerning equipment lifetimes. This allows for not only an average efficiency to be calculated, but also for equipment retirements to be vintaged—older equipment tends to be lower in efficiency and also tends to get retired before newer, more efficient equipment. Once equipment is retired, the Appliance Stock and Technology Choice Modules determine the efficiency of the replacement equipment. It is often the case that the retired equipment is replaced by substantially more efficient equipment.

As the stock efficiency changes over the simulation interval, energy consumption decreases in inverse proportion to efficiency. Also, as efficiency increases, the efficiency rebound effect (discussed below) will offset some of the reductions in energy consumption by increased demand for the end-use service. For example, if the stock average for electric heat pumps is now 10 percent more efficient than in 1997, then all else constant (weather, real energy prices, shell efficiency, etc...), energy consumption per heat pump would average about only 9 percent less.

Adjusting for the Size of New Construction

Information derived from RECS 1997 indicates that new construction (post-1980) is on average roughly 17 percent larger than the existing stock of housing. The residential module uses similar estimates for each Census Division to model the size of new construction by housing type. The energy consumption for space heating, air conditioning, and lighting are assumed to increase with the square footage of the structure (all future new construction is assumed to be of the size of the post-1980 vintage stock from RECS and Bureau Census data[10]). This results in an increase in the average size of the housing stock of 1,663 to 1,707 square feet from 1997 through 2020.

Adjusting for Weather and Climate

Weather in any given year always includes short-term deviations from the expected longer-term average (or climate). Recognition of the effect of weather on space heating and air conditioning is necessary to avoid inadvertently projecting abnormal weather conditions into the future. In the residential module, proportionate adjustments are made to space heating and air conditioning UECs by Census Division. These heating and cooling degree-days (HDD and CDD). A 10 percent increase in HDD would increase space heating consumption by 10 percent over what it would have other wise been. The residential module makes weather adjustments for the years 1997 through 1999. After 1999, long term weather patterns are assumed to occur. The residential module uses 30-year averages of HDD and CDD as normal weather conditions.

Short-Term Price Effect and Efficiency Rebound

It is assumed that energy consumption for a given end-use service is affected by the marginal cost of providing that service. That is, all else equal, a change in the price of a fuel will have an opposite, but less than proportional, effect on fuel consumption. The current value for the short-term elasticity parameter is -0.25. This value implies that for a 1 percent increase in the price of a fuel, there will be a corresponding decrease in energy consumption of -0.25 percent. Another way of affecting the marginal cost of providing a service is through altered equipment efficiency. For example, a 10 percent increase in efficiency will reduce the cost of providing the end-use service by 10 percent. Based on the short-term efficiency rebound parameter, the demand for the service will rise by 1.5 percent (-10 percent multiplied by -0.15). Only space heating and cooling are assumed to be affected by both elasticities and the efficiency rebound effect.

Shell Efficiency

The shell integrity of the building envelope is an important determinant of the heating and cooling load for each type of household. In the NEMS Residential Demand Module, the shell integrity is represented by an index, which changes over time to reflect improvements in the building shell. The shell integrity index is dimensioned by vintage of house, type of house, fuel type, service (heating and cooling), and Census Division. The age, type, location, and type of heating fuel are important factors in determining the level of shell integrity. Housing units which heat with electricity tend to be better insulated than homes that use other fuels. The age of homes are classified by new (post-1997) and existing. Existing homes are characterized by the RECS 1997 survey and are assigned a shell index value based on the mix of homes that exist in the base year (1997). The improvement over time in the shell integrity of these homes is a function of two factors—an assumed annual efficiency improvement and improvements made when real fuel prices increase (no price-related adjustment is made when fuel prices fall). New homes are more efficient than old homes in terms of their building envelope. Based on RECS data and existing building codes, newer homes are roughly 10 percent more efficient than the existing stock, depending upon the heating fuel, housing type, and Census Division. Over time, the shell integrity of new homes improves as the stringency of building codes increases. The shell integrity index affects the space heating and cooling loads directly, causing a decrease in fuel consumed for these services as the shell integrity improves.

Legislation and Other Federal Programs

Energy Policy Act of 1992 (EPACT)

The EPACT contains several policies which are designed to improve residential sector energy efficiency. The EPACT policies analyzed in the NEMS Residential Demand Module include the sections relating to window labeling programs, low-flow showerheads, and building codes. The impact of building codes is captured in the shell efficiency index for new buildings listed above. Other EPACT provisions, such as home energy efficiency ratings and energy-efficient mortgages, which allow home buyers to qualify for higher loan amounts if the home is energy-efficient, are voluntary, and their effects on residential energy consumption have not been estimated.

The window labeling program is designed to help consumers determine which windows are most energy efficient. These labels already exist for all major residential appliances. Based on analysis of RECS data, it is assumed that the window labeling program will decrease heating loads by 8 percent and cooling loads by 3 percent. Approximately 25 percent of the existing (pre-1998) housing stock is affected by this policy by 2015.

The low-flow showerhead program is designed to cut domestic hot water use for showers. It is assumed that these showerheads cut hot water use by 33 percent for shower use. Since showers account for approximately 30 percent of domestic hot water use, total hot water use decreases by 15 percent. It is further assumed that these showerheads are installed exclusively in new construction.

National Appliance Energy Conservation Act of 1987

The Technology Choice Submodule incorporates equipment standards established by the National Appliance Energy Conservation Act of 1987 (NAECA). Some of the NAECA standards implemented in the module include: a Seasonal Energy Efficiency Rating (SEER) of 10.0 for heat pumps; an Annual Fuel Utilization Efficiency (energy output over energy input) of 0.78 for oil and gas furnaces; an Efficiency Factor of .88 for electric water heaters; and refrigerator standards that set consumption limits to 976 kilowatt-hours per year in 1990, 691 kilowatt-hours per year in 1993, and 483 kilowatt-hours per year in 2002.

Climate Change Action Plan

The Climate Change Action Plan (CCAP) contains many policies which are designed to reduce carbon emissions in the United States to the 1990 levels. The CCAP strategies which directly affect the residential sector are Actions 8 through 11. The Residential Demand Module for *AEO2000* includes effects from Action Items 6, 7, 8, 10, and 11 (the House and Senate appropriations included no funding for Action 9). Specifically, these sections relate to Federal Efficiency Standards for several household appliances, stricter building codes, and the expansion of "Golden Carrot" demand-pull type programs. Analyses relating to CCAP programs are on an ongoing basis, as funding changes over time.

Action Item 6 includes voluntary programs sponsored by the Department of energy (DOE) and the Environmental Protection Agency (EPA) aimed at market-pull partnerships with industry. Among the programs in Action Item 6 are DOE's R&D efforts to commercialize advanced energy-efficient technologies and EPA's Energy Star Programs for residential homes, air conditioning, ductwork and lighting.

CCAP Action Items 8, 10 and 11 are policies designed to reduce energy consumption by strengthening building shell efficiency and promoting energy efficient mortgages. In *AEO2000*, the shell integrity (efficiency) of new construction increases relative to 1997 levels as stricter building codes, energy-efficient mortgages, and home energy rating systems become more widespread. The combined energy savings due to CCAP Actions 6 through 11 results in approximately 5.6 MMT of carbon emissions savings in the year 2010 and 13.2 MMT in 2020.

Residential Technology Cases

In addition to the *AEO2000* reference case, three side cases were developed to examine the effect of equipment and building standards on residential energy use—a *2000 technology case,* a *best available technology case*, and a *high technology case*. These side cases were analyzed in stand-alone (not integrated with the supply modules) NEMS runs and thus do not include supply-responses to the altered residential consumption patterns of the two cases. *AEO2000* also analyzed an integrated *high technology case (consumption high technology)*, which combines the *high technology cases* of the four end-use demand sectors and the *electricity high fossil technology case*.

The 2000 technology case assumes that all future equipment purchases are made based only on equipment available in 2000. This case further assumes that building shell efficiencies will not improve beyond 2000 levels. In the reference case, the 2020 housing stock shell efficiency is 10 percent higher than in 1997 for heating (9 percent for cooling).

The *high technology case* assumes earlier availability, lower costs, and/or higher efficiencies for more advanced equipment than the reference case. Equipment assumptions were developed by engineering technology experts, considering the potential impact on technology given increased research and development into more advanced technologies.[11] In the *high technology case*, heating shell efficiency increases by 25 percent and cooling shell efficiency by 23 percent, relative to 1997.

The *best available technology case* assumes that all equipment purchases from 2000 forward are based on the highest available efficiency in the *high technology case* in a particular simulation year, disregarding the economic costs of such a case. It is merely designed to show how much the choice of the highest-efficiency equipment could affect energy consumption. In this case, heating shell efficiency increases by 25 percent and cooling shell efficiency by 23 percent, relative to 1997.

Notes and Sources

[8] The Model Documentation Report contains additional details concerning model structure and operation. Refer to Energy Information Administration, *Model Documentation Report: Residential Sector Demand Module of the National Energy Modeling System,* DOE/EIA-M065(2000), (December 1999).

[9] Among the explanations often mentioned for observed high average implicit discount rates are: market failures, (i.e., cases where incentives are not properly aligned for markets to result in purchases based on energy economics alone); unmeasured technology costs (i.e., extra costs of adoption which are not included or difficult to measure like employee down-time); characteristics of efficient technologies viewed as less desirable than their less efficient alternatives (such as equipment noise levels or lighting quality characteristics); and the risk inherent in making irreversible investment decisions. Examples of market failures/barriers include: decision makers having less than complete information, cases where energy equipment decisions are made by parties not responsible for energy bills (e.g., landlord/tenants, builders/home buyers), discount horizons which are truncated (which might be caused by mean occupancy times that are less than the simple payback time and that could possibly be classified as an information failure), and lack of appropriate credit vehicles for making efficiency investments, to name a few. The use of high implicit discount rates in NEMS merely recognizes that such rates are typically found to apply to energy-efficiency investments.

[10] U.S. Bureau of Census, *Characteristics of New Housing*, C25/95-A.

[11] The high technology assumptions are based on Energy Information Administration, Technology Forecast Updates-Residential and Commercial Building technologies-Advanced Adoption Case (Arthur D. Little, Inc., September 1998).

Commercial Demand Module

The NEMS Commercial Sector Demand Module generates forecasts of commercial sector energy demand through 2020. The definition of the commercial sector is consistent with EIA's State Energy Data System (SEDS). That is, the commercial sector includes business establishments that are not engaged in transportation or in manufacturing or other types of industrial activity (e.g., agriculture, mining or construction). The bulk of commercial sector energy is consumed within buildings; however, street lights, pumps, bridges, and public services are also included if the establishment operating them is considered commercial. Since most of commercial energy consumption occurs in buildings, the commercial module relies on the data from the EIA Commercial Buildings Energy Consumption Survey (CBECS) for characterizing the commercial sector activity mix as well as the equipment stock and fuels consumed to provide end use services.[12]

The commercial module forecasts consumption by fuel[13] at the Census Division level using prices from the NEMS energy supply modules, macroeconomic variables from the NEMS Macroeconomic Activity Module (MAM), as well as external data sources (technology characterizations, for example). Energy demands are forecast for ten end-use services[14] for eleven building categories[15] in each of the nine Census Divisions. The model begins by developing forecasts of floorspace for the 99 building category and Census Division combinations. Next, the ten end-use service demands required for the projected floorspace are developed. Technologies are then chosen to meet the projected service demands for the seven major end uses.[16] Once technologies are chosen, the energy consumed by the equipment stock (both previously existing and purchased equipment) chosen to meet the projected end-use service demands is developed.[17]

Key Assumptions

The key assumptions made by the commercial module are presented in terms of the flow of the calculations described above. Each section below will summarize the assumptions in each of the commercial module submodules: floorspace, service demand, technology choice, and end-use consumption. The four submodules are executed sequentially in the order presented, and the outputs of each submodule become the inputs to subsequently executed submodules. As a result, key forecast drivers for the floorspace submodule are also key drivers for the service demand submodule, and so on.

Floorspace Submodule

Floorspace is forecast by starting with the previous year's stock of floorspace and eliminating a certain portion to represent the age-related removal of buildings. Total floorspace is the sum of the surviving floorspace plus new additions to the stock derived from the Macroeconomic Activity Module's floorspace projection.[18]

Existing Floorspace and Attrition

Existing floorspace is based on the estimated floorspace reported in the *Commercial Buildings Energy Consumption Survey 1995* (Table 10). Over time, the 1995 stock is projected to decline as buildings are removed from service (floorspace attrition). Floorspace attrition is estimated by a logistic decay function, the shape of which is dependent upon the values of two parameters: average building lifetime and *gamma*. The average building lifetime refers to the median expected lifetime of a particular building type. The *gamma* parameter corresponds to the rate at which buildings retire near their median expected lifetime. The current values for the average building lifetime and *gamma* are 59 years and 5.4, respectively.[19]

New Construction Additions to Floorspace

The commercial module develops estimates of projected commercial floorspace additions by combining the surviving floorspace estimates with the Data Resources, Inc. (DRI) total floorspace forecast from MAM. A total NEMS floorspace projection is calculated by applying DRI's assumed floorspace growth rate within each Census Division and DRI building type to the corresponding NEMS Commercial Demand Module's building types based on the CBECS building types shares. The NEMS surviving floorspace from the previous year is then subtracted from the total NEMS floorspace projection for the current year to yield new floorspace additions.[20]

Table 10. 1995 Total Floorspace by Census Division and Principal Building Activity
(Millions of Square Feet)

	Assem-bly	Educa-tion	Food Sales	Food Service	Health Care	Lodging	Large Office	Small Office	Merc/ Service	Ware-house	Other	Total
New England	290	567	11	38	70	150	211	351	820	308	324	3,140
Middle Atlantic	846	1,363	68	127	248	199	1,026	656	2,019	1,172	1,020	8,743
East North Central	1,028	1,336	43	417	250	642	869	747	1,994	1,624	705	9,655
West North Central	563	661	25	57	155	267	358	426	1,209	420	528	4,669
South Atlantic	906	932	107	173	270	729	1,099	1,045	2,103	1,543	568	9,475
East South Central	670	379	50	105	137	324	260	335	1,325	1,032	300	4,917
West South Central	797	1,004	129	164	208	261	482	563	1,436	861	533	6,438
Mountain	707	547	85	58	87	383	435	411	456	522	164	3,855
Pacific	934	951	124	213	217	663	1,016	881	1,366	999	516	7,881
United States	6,741	7,740	642	1,352	1,642	3,618	5,756	5,414	12,728	8,481	4,658	58,772

Source: Energy Information Administration, Commercial Buildings Energy Consumption Survey 1995 Public Use Data.

Note: totals may not equal sum of components due to independent rounding.

Service Demand Submodule

Once the building stock is projected, the Commercial Demand module develops a forecast of demand for energy-consuming services required for the projected floorspace. The module projects service demands for the following explicit end-use services: space heating, space cooling, ventilation, water heating, lighting, cooking, refrigeration, personal computer office equipment, and other office equipment.[21] The service demand intensity (SDI) is measured in thousand Btu of end-use service demand per square foot and differs across service, Census Division and building type. The SDIs are based on a hybrid engineering and statistical approach of CBECS consumption data.[22] Projected service demand is the product of square feet and SDI for all end uses across the eleven building categories with adjustments for changes in shell efficiency for space heating and cooling.

Shell Efficiency

The shell integrity of the building envelope is an important determinant of the heating and cooling loads for each type of building. In the NEMS Commercial Demand Module, the shell efficiency is represented by an index, which changes over time to reflect improvements in the building shell. This index is dimensioned by building type and Census Division and applies directly to heating. For cooling, the effects are computed from the index, but differ from heating effects, because of different marginal effects of shell integrity and because of internal building loads. In the *AEO2000* reference case, shell improvements for new buildings are up to 24 percent more efficient than the 1995 stock of similar buildings. Over the forecast horizon, new building shells improve in efficiency by 6 percent relative to their efficiency in 1995. For existing buildings, efficiency is assumed to increase by 4 percent over the 1995 stock average. The shell efficiency index affects the space heating and cooling service demand intensities causing changes in fuel consumed for these services as the shell integrity improves.

Technology Choice Submodule

The technology choice submodule develops projections of the results of the capital purchase decisions for equipment fueled by the three major fuels (electricity, natural gas, and distillate fuel). Capital purchase decisions are driven by assumptions concerning behavioral rule proportions and time preferences, described below, as well as projected fuel prices, average utilization of equipment (the "capacity factors"), relative technology capital costs, and operating and maintenance (O&M) costs.

Decision Types

In each forecast year, equipment is potentially purchased for three "decision types". Equipment must be purchased for newly added floorspace and to replace a portion of equipment in existing floorspace projected to wear out.[23] Equipment is also potentially purchased for retrofitting equipment which has become economically obsolete. The purchase of retrofit equipment occurs only if the annual operating costs of a current technology exceed the annualized capital and operating costs of a technology available as a retrofit candidate.

Behavioral Rules

The commercial module allows the use of three alternate assumptions about equipment choice behavior. These assumptions constrain the equipment selections to three choice sets, which are progressively more restrictive. The choice sets vary by decision type and building type:

- **Unrestricted Choice Behavior -** This rule assumes that commercial consumers consider *all* types of equipment that meet a given service, across all fuels, when faced with a capital purchase decision.

- **Same Fuel Behavior -** This rule restricts the capital purchase decision to the set of technologies that consume the *same fuel that currently meets the decision maker's service demand.*

- **Same Technology Behavior -** Under this rule, commercial consumers consider only the available models of the *same technology and fuel* that currently meet service demand, when facing a capital stock decision.

Under any of the above three behavior rules, equipment that meets the service at the lowest annualized lifecycle cost is chosen. Table 11 illustrates the proportions of floorspace subject to the different behavior rules for space heating technology choices in large office buildings.

Table 11. Assumed Behavior Rules for Choosing Space Heating Equipment in Large Office Buildings (Percent)

	Unrestricted	Same Fuel	Same Technology	Total
New Equipment Decision	21	30	49	100
Replacement Decision	8	35	57	100
Retrofit Decision	0	5	95	100

Source: Energy Information Administration, *Model Documentation Report: Commercial Sector Demand Module of the National Energy Modeling System*, DOE/EIA-M066(2000) (December 1999).

Time Preferences

The time preferences of owners of commercial buildings are assumed to be distributed among seven alternate time preference premiums (Table 12). Adding the time preference premiums to the 10-year Treasury Bill rate results in implicit discount rates, also known as hurdle rates, applicable to the assumed proportions of commercial floorspace. The effect of the use of this distribution of discount rates is to prevent a single technology from dominating purchase decisions in the lifecycle cost comparisons. The distribution used for *AEO2000* assigns some floorspace a very high discount or hurdle rate to simulate floorspace which will never retrofit existing equipment and which will only purchase equipment with the lowest capital cost. Discount rates for the remaining six segments of the distribution get progressively lower, simulating increased sensitivity to the fuel costs of the equipment that is purchased. The proportion of floorspace assumed for the 0.0 time preference premium represents an estimate of the Federally owned commercial floorspace that is subject to purchase decisions in a given year. In accordance with Executive Order 13123 signed in June 1999, the Federal sector uses a rate comparable to the 10-year Treasury Bill rate when making purchase decisions.

Table 12. Assumed Distribution of Time Preference Premiums
(Percent)

Proportion of Floorspace-All Services Except Lighting	Proportion of Floorspace-Lighting	Time Preference Premium
27.0	27.0	1000.0
25.4	25.4	152.9
20.4	20.4	55.4
16.2	16.2	30.9
10.0	6.0	19.9
0.8	4.8	13.6
0.2	0.2	0.0
100.0	100.0	--

Source: Energy Information Administration. *Model Documentation Report: Commercial Sector Demand Module of the National Energy Modeling System*, DOE/EIA-M066(2000) (Forthcoming, December 1999).

The distribution of hurdle rates used in the commercial module is also affected by changes in fuel prices. If a fuel's price rises relative to its price in the base year (1995), the nonfinancial portion of each hurdle rate in the distribution decreases to reflect an increase in the relative importance of fuel costs, expected in an environment of rising prices. Parameter assumptions for *AEO2000* result in a 30 percent reduction in the nonfinancial portion of a hurdle rate if the fuel price doubles. If the time preference premium input by the model user results in a hurdle rate below the assumed financial discount rate for the commercial sector, 15 percent, with base year fuel prices (such as the rate given in Table 12 for the Federal sector), no response to increasing fuel prices is assumed.

Technology Characterization Database

The technology characterization database organizes all relevant technology data by end use, fuel, and Census Division. Equipment is identified in the database by a technology index as well as a vintage index, the index of the fuel it consumes, the index of the service it provides, its initial market share, the Census Division index for which the entry under consideration applies, its efficiency (or coefficient of performance or efficacy in the case of lighting equipment), installed capital cost per unit of service demand satisfied, operating and maintenance cost per unit of service demand satisfied, average service life, year of initial availability, and last year available for purchase. Equipment may only be selected to satisfy service demand if the year in which the decision is made falls within the window of availability. Equipment acquired prior to the lapse of its availability continues to be treated as part of the existing stock and is subject to replacement or retrofitting. This flexibility in limiting equipment availability allows the direct modeling of equipment efficiency standards. Table 13 provides a sample of the technology data for space heating in the New England Census Division.

Starting with *AEO2000*, an option to allow endogenous price-induced technological change has been included in the determination of equipment costs and availability for the menu of equipment. This concept allows future technologies faster diffusion into the market place if fuel prices increase markedly for a sustained period of time. Although no price-induced change would have been expected using *AEO2000* reference case fuel prices, the option was not exercised for the *AEO2000* model runs.

End-Use Consumption Submodule

The end-use consumption submodule calculates the consumption of each of the three major fuels for the ten end-use services plus fuel consumption for Cogeneration and district services. For the ten end-use services, energy consumption is calculated as the end-use service demand met by a particular type of

Table 13. Capital Cost and Efficiency Ratings of Selected Commercial Space Heating Equipment[1]

Equipment Type	Vintage	Efficiency[2]	Capital Cost ($1987 per Mbtu/hour) [3]	Maintenance Cost ($1987 per Mbtu/hour) [3]	Service Life (Years)
Electric Heat Pump	Current Standard	6.8	$71.92	$2.10	12
	1998- typical	7.5	$77.18	$2.10	12
	1998- high efficiency	9.4	$96.47	$2.10	12
	2005- typical	8.0	$77.18	$2.10	12
	2005- high efficiency	9.5	$94.72	$2.10	12
	2015 - typical	8.5	$73.67	$2.10	12
	2015 - high efficiency	10.0	$91.21	$2.10	12
Ground-Source Heat Pump	1998- typical	3.4	$166.67	$1.35	20
	1998- high efficiency	4.0	$250.00	$1.35	20
	2005- typical	3.4	$145.83	$1.35	20
	2005- high efficiency	4.1	$225.00	$1.35	20
	2015- typical	3.8	$135.42	$1.35	20
	2015 -high efficiency	4.2	$197.92	$1.35	20
Electric Boiler	Current Standard	0.98	$16.48	$0.09	21
Packaged Electric	1995	0.93	$18.63	$3.29	18
Natural Gas Furnace	Current Standard	0.80	$9.21	$0.69	20
	1998- high efficiency	0.92	$11.12	$0.67	20
	2015 - typical	0.81	$9.21	$0.68	20
Natural Gas Boiler	Current Standard	0.80	$7.95	$0.26	25
	1998 - high efficiency	0.90	$11.49	$0.35	25
	2005- typical	0.81	$7.76	$0.26	25
	2005- high efficiency	0.90	$9.49	$0.30	25
Natural Gas Heat Pump	1998- engine driven	4.1	$229.17	$4.69	13
	2005- engine driven	4.1	$166.67	$3.65	13
	2005- absorption	1.4	$173.61	$4.17	15
Distillate Oil Furnace	Current Standard	0.81	$10.58	$0.69	15
	1998	0.83	$16.06	$0.69	15
	2000	0.86	$16.26	$0.69	15
	2010	0.89	$16.81	$0.69	15
Distillate Oil Boiler	Current Standard	0.83	$12.28	$0.06	20
	1998- high efficiency	0.87	$17.19	$0.06	20
	2005- typical	0.83	$12.16	$0.06	20
	2005- high efficiency	0.87	$16.45	$0.06	20

[1]Equipment listed is for the New England Census Division, but is also representative of the technology data for the rest of the U.S.

[2]Efficiency measurements vary by equipment type. Electric air-source and natural gas heat pumps are rated for heating performance using the Heating Seasonal Performance Factor (HSPF); natural gas and distillate furnaces are based on Annual Fuel Utilization Efficiency; ground-source heat pumps are rated on coefficient of performance; and boilers are based on combustion efficiency.

[3]Capital and maintenance costs are given in 1987 dollars.

equipment divided by its efficiency and summed over all existing equipment types. This calculation includes dimensions for Census Division, building type and fuel. Consumption of the five minor fuels is forecast based on historical trends.

Equipment Efficiency

The average energy consumption of a particular appliance is based initially on estimates derived from CBECS 1995. As the stock efficiency changes over the model simulation, energy consumption decreases nearly, but not quite proportionally to the efficiency increase. The difference is due to the calculation of efficiency using the harmonic average and also the efficiency rebound effect discussed below. For example, if on average, electric heat pumps are now 10 percent more efficient than in 1995, then all else constant (weather, real energy prices, shell efficiency, etc...), energy consumption per heat pump would now average about 9 percent less. The Service Demand and Technology Choice Submodules together determine the average efficiency of the stocks used in adjusting the initial average energy consumption.

Adjusting for Weather and Climate

Weather in any given year always includes short-term deviations from the expected longer-term average (or climate). Recognition of the effect of weather on space heating and air conditioning is necessary to avoid projecting abnormal weather conditions into the future. In the commercial module, proportionate adjustments are made to space heating and air conditioning demand by Census Division. These adjustments are based on NOAA data for HDD and CDD. A 10 percent increase in HDD would increase space heating consumption by 10 percent over what it would have been otherwise. The commercial module makes weather adjustments for the years 1996 through 1999. After 1999, long term weather patterns are assumed based on 30-year averages of HDD and CDD.

Short-Term Price Effect and Efficiency Rebound

It is assumed that energy consumption for a given end-use service is affected by the marginal cost of providing that service. That is, all else equal, a change in the price of a fuel will have an inverse, but less than proportional, effect on fuel consumption. The current value for the short-term price elasticity parameter is -0.25 for all end uses except refrigeration. A value of -0.1 is currently used for commercial refrigeration. For example, for lighting this value implies that for a 1 percent increase in the price of a fuel, there will be a corresponding decrease in energy consumption of 0.25 percent. Another way of affecting the marginal cost of providing a service is through equipment efficiency. As equipment efficiency changes over time, so will the marginal cost of providing the end-use service. For example, a 10 percent increase in efficiency will reduce the cost of providing the service by 10 percent. The short-term elasticity parameter for efficiency rebound effects is -0.15 for affected end uses; therefore, the demand for the service will rise by 1.5 percent (-10 percent x -0.15). Currently, all services are affected by the short-term price effect and services affected by efficiency rebound are space heating and cooling, water heating, ventilation and lighting.

Cogeneration

Nonutility power production applications within the commercial sector are concentrated in education, health care, office and warehouse buildings. Program driven installations of solar photovoltaic systems are based on information from DOE's Photovoltaic and Million Solar Roofs programs as well as DOE news releases and the Utility PhotoVoltaic Group web site. Historical data from Form EIA-867, *Annual Nonutility Power Producer Report*, are used to derive electricity cogeneration for 1996 by Census Division, building type and fuel. After 1996, a forecast of distributed generation and cogeneration of electricity is developed based on the economic returns projected for distributed generation and cogeneration technologies. The model uses a detailed cash-flow approach to estimate the number of years required to achieve a cumulative positive cash flow (some technologies may never achieve a cumulative positive cash flow). Penetration assumptions for distributed generation and cogeneration technologies are a function of the estimated number of years required to achieve a positive cash flow.

Legislation and Other Federal Programs

Energy Policy Act of 1992 (EPACT)

A key assumption incorporated in the technology selection process is that the equipment efficiency standards described in the EPACT constrain minimum equipment efficiencies. The effects of standards are modeled by modifying the technology database to eliminate equipment that no longer meets minimum efficiency requirements. For standards effective January 1, 1994, affected equipment includes electric heat pumps—minimum coefficient of performance of 1.64, furnaces and boilers—minimum annual fuel utilization efficiency of 0.8, fluorescent lighting—minimum efficacy of 75 lumens per watt, incandescent lighting—minimum efficacy of 16.9 lumens per watt, air conditioners—minimum seasonal energy efficiency ratio of 10.5, electric water heaters—minimum energy factor of 0.85, and gas and oil water heaters—minimum energy factor of 0.78.

Climate Change Action Plan

The Climate Change Action Plan (CCAP) contains 5 Action Items which affect the commercial sector. Action Items 1, 4 and 5 are designed to stimulate investment in more efficient building shells and equipment for heating, cooling and other end uses. Action Item 2, EPA's Green Lights Program targets the retrofitting of lighting equipment. Action Item 3 was unfunded and therefore not modeled. The commercial module includes several features that allow projected efficiency to increase in response to voluntary programs (e.g., the distribution of time preference premiums and shell efficiency parameters). For Action Items 1, 2, 4 and 5, retrofits of equipment for space heating and air conditioning are incorporated in the distribution of premiums given in Table 11. Also, based partly on these actions, the shell efficiency of new and existing buildings is assumed to increase from 1995 through 2020. Shells for new buildings increase in efficiency by 6 percent over this period, while shells for existing buildings increase in efficiency by 4 percent. In total, the action tems result in energy savings which are estimated to reduce carbon emissions by the commercial sector by 11.5 million metric tons for the year 2010.

Commercial Technology Cases

In addition to the *AEO2000* reference case, three side cases were developed to examine the effect of equipment and building standards on commercial energy use—a 2000 technology case, a *high technology case*, and a *best available technology case*. These side cases were analyzed in stand-alone (not integrated with the NEMS demand and supply modules) commercial model runs and thus do not include supply-responses to the altered commercial consumption patterns of the three cases. *AEO2000* also analyzed an integrated high technology case (*consumption high technology*), which combines the *high technology cases* of the four end-use demand sectors and the *electricity high fossil technology case*.

The *2000 technology case* assumes that all future equipment purchases are made based only on equipment available in 2000. This case further assumes building shell efficiency to be fixed at 2000 levels. In the reference case, existing building shells are allowed to increase in efficiency by 4 percent over 1995 levels, new building shells improve by 6 percent by 2020 relative to new buildings in 1995.

The *high technology case* assumes earlier availability, lower costs, and/or higher efficiencies for more advanced equipment than the reference case. Equipment assumptions were developed by engineering technology experts, considering the potential impact on technology given increased research and development into more advanced technologies. In the *high technology case*, building shell efficiencies are assumed to improve 50 percent faster than in the *reference case* from 2000 forward. Existing building shells, therfore, increase by 5.6 percent relative to 1995 levels and new building shells by 8.4 percent relative to their efficiency in 1995 by 2020.

The *best available technology case* assumes that all equipment purchases from 2000 forward are based on the highest available efficiency in the *high technology case* in a particular simulation year, disregarding the economic costs of such a case. It is merely designed to show how much the choice of the highest-efficiency equipment could affect energy consumption. Shell effects in this case are assumed to be the same as for the *high technology case* above.

Fuel shares, where appropriate for a given end use, are allowed to change in the technology cases as the available technologies from each technology type compete to serve certain segments of the commercial floorspace market. For example, in the *best available technology case*, the most efficient gas furnace technology competes with the most efficient electric heat pump technology. This contrasts with the reference case, in which, a greater number of technologies for each fuel with varying efficiencies all compete to serve the heating end use. In general, the fuel choice will be affected as the available choices are constrained or expanded, and will thus differ across the cases.

Buildings Standards Cases

The buildings sector includes two cases to examine the potential effects of future appliance efficiency standards on energy consumption. For these cases, near-term efficiency standards and the effective date of the standard are based on the American Council for an Energy Efficient Economy's *Approaching the Kyoto Targets: Five Key Strategies for the United States*. Future updates to these standards are assumed to occur every eight years, increasing the efficiency level by 10 and 20 percent in the two cases, respectively, if technically feasible.

Notes and Sources

[12] Energy Information Administration, A Look at Commercial Buildings in 1995: Characteristics, Energy Consumption, and Energy Expenditures, DOE/EIA-0625(95), (Washington, DC, October 1998).

[13] The fuels accounted for by the commercial module are electricity, natural gas, distillate fuel oil, residual fuel oil, liquefied petroleum gas (LPG), coal, motor gasoline, and kerosene. In addition to these fuels the use of solar energy is projected based on an exogenous forecast of projected solar photovoltaic system installations under the Million Solar Roofs program and the potential endogenous penetration of solar photovoltaic systems and solar thermal water heaters.

[14] The end-use services in the commercial module are heating, cooling, water heating, ventilation, cooking, lighting, refrigeration, PC and non-PC office equipment and a category denoted other to account for all other minor end uses.

[15] The 11 building categories are assembly, education, food sales, food services, health care, lodging, large offices, small offices, mercantile/services, warehouse and other.

[16] Minor end uses are modeled based on penetration rates and efficiency trends.

[17] The detailed documentation of the commercial module contains additional details concerning model structure and operation. Refer to Energy Information Administration, Model Documentation Report: Commercial Sector Demand Module of the National Energy Modeling System, DOE/EIA M066(2000), (December 1999).

[18] The floorspace from the Macroeconomic Activity Model is based on the Data Resources Incorporated (DRI) floorspace estimates which are approximately 15 percent lower than the estimate obtained from the CBECS used for the Commercial module. The DRI forecast is developed using the F.W. Dodge data on commercial floorspace. See F.W. Dodge, Building Stock Database Methodology and 1991 Results, Construction Statistics and Forecasts, F.W. Dodge, McGraw-Hill.

[19] The commercial module performs attrition for 9 vintages of floorspace developed from the CBECS 1995 stock estimate and historical floorspace additions data from F.W. Dodge data.

[20] In the event that the computation of additions produce a negative value for a specific building type, it is assumed to be zero.

[21] "Other office equipment" includes copiers, fax machines, typewriters, cash registers, and other miscellaneous office equipment. A tenth category denoted other includes equipment such as elevators, medical, and other laboratory equipment, communications equipment, security equipment, and miscellaneous electrical appliances. Commercial energy consumed outside of buildings and for cogeneration is also included in the "other" category.

[22] Based on updated estimates using CBECS 1995 data and the methodology described in End-Use Energy Consumption Estimates for U.S. Commercial Buildings, 1992, Belzer, D.B., and Wrench,

Notes and Sources

L.E., Pacific Northwest Laboratories, PNNL-11514, Prepared for the U.S. DOE under Contract DE-AC06-76RLO-1830, (Richland, WA, March, 1997).

[23] The proportion of equipment retiring is inversely related to the equipment life.

Industrial Demand Module

The NEMS Industrial Demand Module estimates energy consumption by energy source (fuels and feedstocks) for 9 manufacturing and 6 nonmanufacturing industries. The manufacturing industries are further subdivided into the energy-intensive manufacturing industries and nonenergy-intensive manufacturing industries. The distinction between the two sets of manufacturing industries pertains to the level of modeling. The energy-intensive industries are modeled through the use of a detailed process flow accounting procedure, whereas the nonenergy-intensive and the nonmanufacturing industries are modeled with substantially less detail (Table 14). The Industrial Demand Module forecasts energy consumption at the four Census region levels; energy consumption at the Census Division level is allocated by using the SEDS[24] data.

Table 14. Industry Categories

Energy-Intensive Manufacturing		Nonenergy-Intensive Manufacturing		Nonmanufacturing Industries	
Food and Kindred Products	(SIC 20)	Metals-Based Durables	(SIC 34, 35, 36, 37, 38)	Agricultural Production -Crops	(SIC 01)
Paper and Allied Products	(SIC 26)	Other Manufacturing	(all remaining manufacturing SIC)	Other Agriculture Including Livestock	(SIC 02, 07, 08, 09)
Bulk Chemicals	(SIC 281, 282, 286, 287)			Coal Mining	(SIC 12)
Glass and Glass Products	(SIC 321, 322, 329)			Oil and Gas Mining	(SIC 13)
Hydraulic Cement	(SIC 324)			Metal and Other Nonmetallic Mining	(SIC 10, 14)
Blast Furnaces and Basic Steel	(SIC 331)			Construction	(SIC 15, 16, 17)
Aluminum	(SIC 3334, 3353)				

SIC = Standard Industrial Classification.

Source: Office of Management and Budget, Standard Industrial Classification Manual 1987 (Springfield, VA, National Technical Information Service).

The energy-intensive industries (food and kindred products, paper and allied products, bulk chemicals, glass and glass products, hydraulic cement, blast furnace and basic steel products, and aluminum) are modeled in considerable detail. Each industry is modeled as three separate but interrelated components consisting of the Process Assembly (PA) Component, the Buildings Component (BLD), and the Boiler/Steam/Cogeneration (BSC) Component. The BSC Component satisfies the steam demand from the PA and BLD Components. In some industries, the PA Component produces byproducts that are consumed in the BSC Component. For the energy-intensive industries, the PA Component is separated into the major production processes or end uses.

Petroleum refining (Standard Industrial Classification 2911) is modeled in detail in the Petroleum Market Module of NEMS, and the projected energy consumption is included in the manufacturing total. Forecasts of refining energy use and oil and gas lease and plant fuel and fuels consumed in cogeneration (Standard Industrial Classification 1311) are exogenous to the Industrial Demand Module, but endogenous to the NEMS modeling system.

Key Assumptions

The NEMS Industrial Demand Module primarily uses a bottom-up process modeling approach. An energy accounting framework traces energy flows from fuels to the industry's output. An important assumption in the development of this system is the use of 1994 baseline Unit Energy Consumption (UEC) estimates based on analysis of the Manufacturing Energy Consumption Survey 1994.[25] The UEC represents the energy required to produce one unit of the industry's output. The output may be defined in terms of physical units (e.g., tons of steel) or in terms of the dollar value of output.

The module depicts the seven most energy-intensive manufacturing industries (apart from petroleum refining, which is modeled in the Petroleum Market Module of NEMS) with a detailed process flow approach. The dominant process technologies are characterized by a combination of unit energy consumption estimates and "technology possibility curves." The technology possibility curves indicate the energy intensity of new and existing stock relative to the 1994 stock over time. Rates of energy efficiency improvements assumed for new and existing plants vary by industry and process. These assumed rates were developed using professional engineering judgments regarding the energy characteristics, year of availability, and rate of market adoption of new process technologies.

Process/Assembly Component

The Process/Assembly (PA) Component models each major manufacturing production step for the energy-intensive industries. The throughput production for each process step is computed as well as the energy required to produce it.

Within this component, the UEC is adjusted based on the technology possibility curves for each step. For example, state-of-the-art additions to waste fiber pulping capacity are assumed to require only 93 percent as much energy as does the average existing plant (Table 15). The technology possibility curve is a means of embodying assumptions regarding new technology adoption in the manufacturing industry and the associated increased energy efficiency of capital without characterizing individual technologies. It is unlikely that new technology is employed in all new capacity additions. Many facilities will only partially incorporate the technology or will need time to debug the operating aspects of the newly installed capacity. For these reasons, it is assumed, that on average, the UEC of additions will fall between the existing UEC and the state-of-the-art UEC. To some extent, all industries will increase the energy efficiency of their process and assembly steps. The reasons for the increased efficiency are not likely to be directly attributable to changing energy prices but due to other exogenous factors. Since the exact nature of the technology improvement is too uncertain to model in detail, the module employs a technology possibility curve to characterize the bundle of technologies available for each process step.

Fuel shares for process and assembly energy use in six of the energy-intensive manufacturing industries[26] are adjusted for changes in relative fuel prices. The six industries are food, paper, chemicals, glass, cement, and steel. In each industry, two logit fuel-sharing equations are applied to revise the initial fuel shares obtained from the process-assembly component. The resharing does not affect the industry's total energy use-only the fuel shares. The methodology adjusts total fuel shares across all process stages and vintages of equipment to account for aggregate market response to changes in relative fuel prices.

The fuel share adjustments are done in two stages. The first stage determines the fuel shares of electricity and nonelectricity energy. (Non-electric energy group excludes boiler fuel and feedstocks.) The second stage determines the fossil fuel shares of nonelectricity energy. In each stage, a new fuel-group share, $NEWSHR_i$, is established as a function of the initial, default fuel-group shares, $DEFLTSHR_j$ and fuel-group prices indices, $PRCRAT_i$. The $DEFLTSHR_i$ are the base year shares. The price indices are the ratio of the current year price to the base year price, in real dollars. The formulation is as follows:

$$NEWSHR_i = \frac{DEFLTSHR_i * e^{\beta_i (1-PRCRAT_i)}}{\sum\limits_{j=1}^{N} DEFLTSHR_i * e^{\beta_j (1-PRCRAT_j)}}$$

The coefficients β_j are all assumed to be 0.2.

Table 15. Coefficients for Technology Possibility Curve

Industry/ Process Unit	Old Facilities		New Facilities		
	REI 2020	TPC	REI 1994	REI 2020	TPC
Food	0.892	-0.0044	0.900	0.792	-0.0049
Pulp & Paper					
Wood Preparation	0.909	-0.0037	0.840	0.830	-0.0004
Waste Pulping	0.938	-0.0025	0.930	0.882	-0.0021
Mechanical Pulping	0.904	-0.0039	0.840	0.821	-0.0009
Semi-Chemical	0.870	-0.0054	0.794	0.756	-0.0019
Kraft, Sulfite, misc. chemicals	0.784	-0.0093	0.730	0.590	-0.0082
Bleaching	0.879	-0.0050	0.852	0.769	-0.0039
Paper Making	0.763	-0.0104	0.750	0.546	-0.0122
Bulk Chemicals	0.892	-0.0044	0.900	0.792	-0.0049
Glass[1]					
Batch Preparation	0.936	-0.0025	0.882	0.882	0
Melting/Refining	0.783	-0.0094	0.877	0.577	-0.0160
Forming	0.912	-0.0035	0.921	0.831	-0.0040
Post-Forming	0.871	-0.0053	0.780	0.759	-0.0011
Cement					
Dry Process	0.815	-0.0078	0.790	0.646	-0.0077
Wet Process[2]	0.954	-0.0025	NA	NA	NA
Finish Grinding	0.899	-0.0041	0.813	0.813	0
Steel[3]					
Coke Oven	0.904	-0.0039	0.840	0.820	-0.0009
BF/BOF	0.899	-0.0041	1.000	0.799	-0.0086
EAF	0.919	-0.0033	0.960	0.841	-0.0051
Ingot Casting/Primary Rolling[2]	1.000	0	NA	NA	NA
Continuous Casting	1.000	0	1.000	1.000	0
Hot Rolling	0.672	-0.0152	0.500	0.381	-0.0104
Cold Rolling	0.768	-0.0101	0.840	0.550	-0.0162
Aluminum					
Primary aluminum	0.898	-0.0041	0.910	0.804	-0.0048
Semi-Fabrication	0.734	-0.0118	0.610	0.497	-0.0078

[1]REIs and TPCs apply to virgin and recycled materials.
[2]No new plants are likely to be built with these technologies.
[3]Net shape casting is projected to reduce the energy requirements for hot and cold rolling rather than for the continuous casting step.
[4]SIC = Standard Industrial Classification.
REI 1994 New Facilities = For new facilities, the ratiio of State-of-the-art energy intensity to average 1994 energy intensity for existing facilities.
REI 2020 Existing Facilities = Ratio of 2020 energy intensity to average 1994 energy intensity for existing facilities.
REI 2020 New Facilities = Ratio of 2020 energy intensity for a new State-of-the-art facility to the average 1994 intensity for existing facilities.
TPC = annual rate of change between 1994 and 2020.
NA = Not applicable.
BF = Blast furnace.
BOF = Basic oxygen furnace.
EAF = Electric arc furnace.
Source: Energy Information Administration, *Model Documentation Report: Industrial Sector Demand Module of the National Energy Modeling System*, DOE/EIA-M064(2000), (Washington, DC, January 2000).

The form of the equation results in unchanged fuel shares when the price indices are all 1, or unchanged from their 1997 levels. The implied own-price elasticity of demand is about -0.1.

Byproducts produced in the PA Component serve as fuels for the BSC Component. In the industrial module, byproducts are assumed to be consumed before purchased fuel.

Buildings Component

The total buildings energy demand by industry for each region is the product of the building UEC and regional industrial employment. Building UEC's were derived by first estimating energy requirements for building lighting, air conditioning, and space heating, where space heating was further divided to estimate the amount provided by direct combustion of fossil fuels and that provided by steam (Table 16). Energy consumption in the BLD Component for an industry is assumed to grow at the same rate as regional employment for that industry.

Table 16. Building Component Unit Energy Consumption
(Trillion Btu/Thousand People Employed)

Industry	Lighting Electric UEC	Building Use and Energy Source		
		HVAC		
		Electric UEC	Natural Gas UEC	Steam UEC
Food & Kindred Products	0.007	0.009	0.014	0.045
Paper & Allied Products	0.0131	0.016	0.023	0.0082
Bulk Chemicals	0.0159	0.0299	0.68	0.0058
Glass and Glass Products	0.0133	0.019	0.044	0.004
Hydraulic Cement	0.029	0.029	0.029	0.0568
Blast Furnaces & Basic Steel	0.0123	0.0184	0.0674	0.011
Primary Aluminum	0.0187	0.0266	0.0062	0.0053
Metal Based Durables	0.0083	0.0125	0.0153	0.0019
Other Non-Intensive MFG Fabricated Metals	0.007	0.0103	0.0134	0.0036

UEC = Unit Energy Consumption.

HVAC = Heating, Ventilation, Air Conditioning.

Source: Energy Information Administration, *Model Documentation Report: Industrial Sector Demand Module of the National Energy Modeling System*, DOE/EIA-M064(2000), (Washington, DC, January 2000).

Boiler/Steam/Cogeneration Component

The steam demand and byproducts from the PA and BLD Components are passed to the BSC Component, which applies a heat rate and a fuel share equation (Table 17) to the boiler steam requirements to compute the required energy consumption.

The boiler fuel shares apply only to the fuels that are used in non-cogeneration boilers. The portion of the steam demand that is met with cogenerated steam reduces the amount of boiler fuel that would otherwise be required. The non-cogeneration boiler fuel shares are calculated using a logit formulation. The equation is calibrated to 1994 so that the actual boiler fuel shares are produced for the relative prices that prevailed in 1994. The equation for each manufacturing industry is as follows:

$$ShareFuel_i = \frac{P_i{}^{\alpha}\beta_i}{\sum\limits_{i=1}^{3} P_i{}^{\alpha}(\beta_i)}$$

where the fuels are coal, petroleum, and natural gas. The P_i are the fuel prices; α a sensitivity parameter; and the β_i are the 1994 fuel shares. The byproduct fuels are consumed before the quantity of purchased fuels is estimated. The boiler fuel shares are based on the 1994 MECS.[27]

Table 17. Logit Function Parameters for Estimating Boiler Fuel Shares

Industry	Alpha	Natural Gas	Steam Coal	Oil
Food	-0.25	0.62	0.16	0.23
Paper and Allied Products	-0.25	0.56	0.31	0.13
Bulk Chemicals	-0.25	0.57	0.18	0.25
Glass and Glass Products	-0.25	0.91	0.0	0.09
Cement	-0.25	0.96	0.02	0.02
Steel	-0.25	0.47	0.15	0.38
Aluminum	-0.25	0.97	0.0	0.03
Metals-Based Durables	-0.25	0.73	0.18	0.10
Other Non-Int MFG	-0.25	0.91	0.05	0.04

Source: Energy Information Administration, *Model Documentation Report: Industrial Sector Demand Module of the National Energy Modeling System*, DOE/EIA-M064(2000), (Washington, DC, January 2000).

Alpha: User-specified.

Cogeneration

Cogeneration (the generation of electricity and steam) has been a standard practice in the industrial sector for many years. The cogeneration estimates in the module are based on the assumption that the historical relationship between industrial steam demand and cogeneration will continue in the future. The data source is Form EIA-867, *Annual Nonutility Power Producer Report*, consisting of data from approximately 400 cogenerators for 1989-1994.

The projection for additions to fossil-fueled cogeneration are determined with a new modeling approach developed for the *Annual Energy Outlook 2000*. The new approach is based on assessing capacity that could be added to generate the industrial steam requirements that are not already met by existing cogeneration. The technical potential for traditional cogeneration is primarily based on supplying thermal requirements. Currently, the approach is based on gas turbine cogeneration plants. Capacity additions are then determined by the interaction of payback periods and market penetration rates.

Technology

The amount of energy consumption reported by the industrial module is also a function of vintage of the capital stock that produces the output. It is assumed that new vintage stock will consist of state-of-the-art technologies that are more energy efficient than the average efficiency of the existing capital stock. Consequently, the amount of energy required to produce a unit of output using new capital stock is less than that required by the existing capital stock. Capital stock is grouped into three vintages: old, middle, and new. The old vintage consists of capital in production prior to 1995 and is assumed to retire at a fixed rate each year (Table 18). Middle vintage capital is that which is added after 1994 but not including the year of the forecast. New production capacity is built in the forecast years when the capacity of the existing stock of capital in the industrial model cannot produce the output forecasted by the NEMS Regional Macroeconomic Model. Capital additions during the forecast horizon are retired in subsequent years at the same rate as the pre-1995 capital stock.

The energy intensity of the new capital stock relative to 1994 capital stock is reflected in the parameter of the technology possibility curve estimated for the major production steps for each of the energy-intensive industries. These curves are based on engineering judgment of the likely future path of energy intensity changes (Table 15). The energy intensity of the existing capital stock also is assumed to decrease over time, but not as rapidly as new capital stock. The net effect is that over time the amount of energy required to produce a unit of output declines. Although total energy consumption in the industrial sector is projected to increase, overall energy intensity is projected to decrease.

Table 18. Retirement Rates

Industry	Retirement Rate (percent)	Industry	Retirement Rate (percent)
Food and Kindred Products.	1.7	Glass and Glass Products	1.3
Pulp and Paper	2.3	Hydraulic Cement	1.2
Bulk Chemicals	1.9	Glass and Glass Products	1.3
Blast Furnace and Basic Steel Products		Primary Aluminum	2.1
Blast Furnace/Basic Oxygen Furnace . .	1.0	Metal-Based Durables.	1.5
Electric Arc Furnace	1.5		
Coke Ovens	1.5		
Other Steel	2.9		

Source: Energy Information Administration, *Model Documentation Report: Industrial Sector Demand Module of the National Energy Modeling System*, DOE/EIA-MO64(2000), (Washington, DC, January 2000).

Note: Except for the Blast Furnace and Basic Steel Products Industry, the retirement rate is the same for each process step or end-use within an industry.

Legislation

Energy Policy Act of 1992 (EPACT)

EPACT and the Clean Air Act Amendments of 1990 (CAAA90) contain several implications for the industrial module. These implications fall into three categories: coke oven standards; efficiency standards for boilers, furnaces, and electric motors; and industrial process technologies. The industrial module assumes the leakage standards for coke oven doors do not reduce the efficiency of producing coke or increase unit energy consumption. The industrial module uses heat rates of 1.25 (80 percent efficiency) and 1.22 (82 percent efficiency) for gas and oil burners respectively. These efficiencies meet the EPACT standards. The standards for electric motors call for a 10-percent efficiency increase. The industrial module incorporates a 10-percent savings for state-of-the-art motors increasing to 20-percent savings in 2015. Given the time lag in the legislation and the expected lifetime of electric motors, no further adjustments are necessary to meet the EPACT standards for electric motors. The industrial module incorporates the necessary reductions in unit energy consumption for the energy-intensive industries.

Climate Change Action Plan

Several programs included in the Climate Change Action Plan (CCAP) target the industrial sector. Note that the potential impacts of the Climate Wise Program are also included in the CCAP impacts. The intent of these programs is to reduce greenhouse gas emissions by lowering industrial energy consumption. The Department of Energy (DOE) program offices estimated that full implementation of these programs would reduce industrial electricity consumption by 79 billion kilowatthours and fossil energy consumption by 359 trillion Btu by 2010. However, since the energy savings associated with the voluntary programs in the CCAP largely duplicate savings that would have occurred in their absence and since some programs were not fully funded, total CCAP energy savings were reduced. The *Annual Energy Outlook 2000* (*AEO2000*) assumes that CCAP reduces electricity consumption by 25 billion kilowatthours and fossil energy consumption by 65 trillion Btu in 2010. The fossil energy is assumed to be 85 percent natural gas and 15 percent steam coal. In this situation, carbon emissions in the industrial sector would be reduced by about 5 million metric tons (1 percent) in 2010.

High Technology and 2000 Technology Cases

The high *technology case* assumes earlier availability, lower costs, and higher efficiency for more advanced equipment.[28] Changes in aggregate energy intensity result both from changing equipment and production efficiency and from changes in the composition of industrial output. Since the composition of industrial output remains the same as in the reference case, aggregate intensity declines by only 1.2 percent annually even though the intensity declines for some individual industries doubles. In the reference case, aggregate intensity declines by 1.0 percent annually.

AEO2000 also analyzed an integrated high technology case (*consumption high technology*), which combines the *high technology cases* of the four end-use demand sectors and the *electricity high fossil technology case*.

The *2000 technology case* holds the energy efficiency of plant and equipment constant at the 2000 level over the forecast. Both cases were run with only the Industrial Demand Module rather than as a fully integrated NEMS run, (i.e., the other demand models and the supply models of NEMS were not executed). Consequently, no potential feedback effects from energy market interactions were captured.

Notes and Sources

[24] Energy Information Administration, *State Energy Data Report 1996, DOE/EIA-0214(96)*, (Washington, D.C., February 1999).

[25] Energy Information Administration, *Manufacturing Consumption of Energy 1994, DOE/EIA-0512(94)*, (Washington, D.C., December 1997).

[26] Aluminum is excluded due to its almost exclusive reliance on electricity in the process and assembly component.

[27] Energy Information Administration, *Manufacturing Consumption of Energy 1994, DOE/EIA-0512(94)*, (Washington, D.C., December 1997).

[28] These assumptions are based in part on Arthur D. Little, "*Aggressive Technology for the NEMS model,*" (September 1998).

Transportation Demand Module

The NEMS Transportation Demand Module estimates energy consumption across the nine Census Divisions and over ten fuel types. Each fuel type is modeled according to fuel-specific technology attributes applicable by transportation mode. Total transportation energy consumption is the sum of energy use in eight transport modes: light-duty vehicles (cars, light trucks, industry sport utility vehicles and vans), commercial light trucks (8501-10,000 lbs), freight trucks (>10,000 lbs), freight and passenger airplanes, freight rail, freight shipping, mass transit, and miscellaneous transport such as mass transit. Light-duty vehicle fuel consumption is further subdivided into personal usage and commercial fleet consumption.

Key Assumptions

Macroeconomic Sector Inputs

Macroeconomic sector inputs used in the NEMS Transportation Demand Module (Table 19) consist of the following: gross domestic product (GDP), industrial output by Standard Industrial Classification code, personal disposable income, new car and light truck sales, total population, driving age population, total value of imports and exports, and the military budget. The share of total vehicle sales that represent light truck sales is assumed to approach fifty-one percent by 2020.

Table 19. Macroeconomic Inputs to the Transportation Module
(Millions)

Macroeconomic Input	1998	2000	2005	2010	2015	2020
New Car Sales	7.7	7.3	7.3	7.2	7.5	7.5
New Light Truck Sales	6.2	6.2	6.9	7.1	7.7	7.7
Real Disposable Income (billion 1992 Chain-Weighted Dollars)	5,348	5,665	6,406	7,204	8,083	9,008
Real GDP (billion 1992 Chain-Weighted Dollars)	7,552	7,991	9,056	10,054	11,147	12,179
Driving Age Population	208.6	212.8	223.7	235.2	245.6	255.3
Total Population	270.6	275.2	286.6	298.3	310.8	323.4

Source: Energy Information Administration, *AEO2000* National Energy Modeling System run: aeo2k.d100199a.

Light-Duty Vehicle Assumptions

The light duty vehicle Fuel Economy Module includes 59 fuel saving technologies with data specific to cars and light trucks including incremental fuel efficiency improvement, incremental cost, first year of introduction, and fractional horsepower change. These assumed technology characterizations are scaled up or down to approximate the differences in each attribute for 6 EPA size classes of cars and light trucks (Tables 20 and 21).

The vehicle sales share module holds vehicle sales shares by import and domestic manufacturers constant within a vehicle size class at the 1998 level from the National Highway Traffic and Safety Administration data.[29]

EPA size class sales shares are projected as a function of income per capita, fuel prices, and average predicted vehicle prices based on endogeous calculations within the Fuel Economy Module.[30]

The fuel economy module utilizes 59 new texchnologies for each size class and origin of manufacturer (domestic or foreign) based on the cost-effectiveness of each technology and an initial availability year. The

Table 20. Standard Technology Matrix For Cars[1]

	Fractional Fuel Efficiency Change	Incremental Cost (1990 $)	Incremental Cost ($/Unit Wt.)	Incremental Weight (Lbs.)	Incremental Weight (Lbs./Unit Wt.)	First Year Introduced	Fractional Horsepower Change
Front Wheel Drive	0.060	160	0.00	0	-0.08	1980	0
Unit Body	0.040	80	0.00	0	-0.05	1980	0
Material Substitution II	0.033	0	0.60	0	-0.05	1987	0
Material Substitution III	0.066	0	0.80	0	-0.10	1997	0
Material Substitution IV	0.099	0	1.00	0	-0.15	2007	0
Material Substitution V	0.099	0	1.50	0	-0.20	2017	0
Drag Reduction II	0.132	32	0.00	0	0.00	1985	0
Drag Reduction III	0.023	64	0.00	0	0.05	1991	0
Drag Reduction IV	0.046	112	0.00	0	0.01	2004	0
Drag Reduction V	0.069	176	0.00	0	0.02	2014	0
TCLU	0.092	40	0.00	0	0.00	1980	0
4-Speed Automatic	0.030	225	0.00	30	0.00	1980	0.05
5-Speed Automatic	0.045	325	0.00	40	0.00	1995	0.07
CVT	0.100	250	0.00	20	0.00	1995	0.07
6-Speed Manual	0.020	100	0.00	30	0.00	1991	0.05
Electronic Transmission I	0.005	20	0.00	5	0.00	1988	0
Electronic Transmission II	0.015	40	0.00	5	0.00	1998	0
Roller Cam	0.020	16	0.00	0	0.00	1987	0
OHC 4	0.030	100	0.00	0	0.00	1980	0.20
OHC 6	0.030	140	0.00	0	0.00	1980	0.20
OHC 8	0.030	170	0.00	0	0.00	1980	0.20
4C/4V	0.080	240	0.00	30	0.00	1988	0.45
6C/4V	0.080	320	0.00	45	0.00	1991	0.45
8C/4V	0.080	400	0.00	60	0.00	1991	0.45
Cylinder Reduction	0.030	-100	0.00	-150	0.00	1988	-0.10
4C/5V	0.100	300	0.00	45	0.00	1998	0.55
Turbo	0.050	500	0.00	80	0.00	1980	0.45
Engine Friction Reduction I	0.020	20	0.00	0	0.00	1987	0
Engine Friction Reduction II	0.035	50	0.00	0	0.00	1996	0
Engine Friction Reduction III	0.050	90	0.00	0	0.00	2006	0
Engine Friction Reduction IV	0.065	140	0.00	0	0.00	2016	0
VVT I	0.080	140	0.00	40	0.00	1998	0.10
VVT II	0.100	180	0.00	40	0.00	2008	0.15
Lean Burn	0.100	150	0.00	0	0.00	2099	0
Two Stroke	0.150	150	0.00	-150	0.00	2099	0
TBI	0.020	40	0.00	0	0.00	1982	0.05
MPI	0.035	80	0.00	0	0.00	1987	0.10
Air Pump	0.010	0	0.00	-10	0.00	1982	0
DFS	0.015	15	0.00	0	0.00	1987	0.10
Oil 5W-30	0.005	2	0.00	0	0.00	1987	0
Oil Synthetic	0.015	5	0.00	0	0.00	1997	0
Tires I	0.010	16	0.00	0	0.00	1992	0
Tires II	0.020	32	0.00	0	0.00	2002	0
Tires III	0.030	48	0.00	0	0.00	2012	0
Tires IV	0.040	64	0.00	0	0.00	2018	0
ACC I	0.005	15	0.00	0	0.00	1992	0
ACC II	0.010	30	0.00	0	0.00	1997	0
EPS	0.015	40	0.00	0	0.00	2002	0
4WD Improvements	0.030	100	0.00	0	-0.05	2002	0
Air Bags	-0.010	300	0.00	35	0.00	1987	0
Emissions Tier I	-0.010	150	0.00	10	0.00	1994	0
Emissions Tier II	-0.010	300	0.00	20	0.00	2003	0
ABS	-0.005	300	0.00	10	0.00	1987	0
Side Impact	-0.005	100	0.00	20	0.00	1996	0
Roof Crush	-0.003	100	0.00	5	0.00	2001	0
Increased Size/Wt.	-0.133	0	0.00	0	0.20	1991	0
GDI/4-cyl	0.170	1000	0.00	0	0.00	2005	0
GDI/6-cyl	0.170	1200	0.00	0	0.00	2005	0
Gasoline Hybrid	0.450	0	75.00	0	0.05	2001	0

N/A = Non Applicable

[1] Fractional changes refer to the percentage change from the 1990 values.

Source: Energy and Environment Analysis, *Changes to the Fuel Economy Module Final Report, prepared for the Energy Information Administration (EIA), (June 1998).*

Table 21. Standard Technology Matrix For Trucks[1]

	Fractional Fuel Efficiency Change	Incremental Cost (1990 $)	Incremental Cost ($/Unit Wt.)	Incremental Weight (Lbs.)	Incremental Weight (Lbs./Unit Wt.)	First Year Introduced	Fractional Horsepower Change
Front Wheel Drive	0.020	160.00	0.00	0	-0.08	1985	0
Unit Body	0.060	80.00	0.00	0	-0.05	1995	0
Material Substitution II	0.033	0.00	0.60	0	-0.05	1996	0
Material Substitution III	0.066	0.00	0.80	0	-0.10	2006	0
Material Substitution IV	0.099	0.00	1.00	0	-0.15	2016	0
Material Substitution V	0.132	0.00	1.50	0	-0.20	2026	0
Drag Reduction II	0.023	32.00	0.00	0	0.00	1990	0
Drag Reduction III	0.046	64.00	0.00	0	0.05	1997	0
Drag Reduction IV	0.069	112.00	0.00	0	0.01	2007	0
Drag Reduction V	0.092	176.00	0.00	0	0.02	2017	0
TCLU	0.030	40.00	0.00	0	0.00	1980	0
4-Speed Automatic	0.045	225.00	0.00	30	0.00	1980	0.05
5-Speed Automatic	0.065	325.00	0.00	40	0.00	1997	0.07
CVT	0.100	250.00	0.00	20	0.00	2005	0.07
6-Speed Manual	0.020	100.00	0.00	30	0.00	1997	0.05
Electronic Transmission I	0.005	20.00	0.00	5	0.00	1991	0
Electronic Transmission II	0.015	40.00	0.00	5	0.00	2006	0
Roller Cam	0.020	16.00	0.00	0	0.00	1986	0
OHC 4	0.030	100.00	0.00	0	0.00	1980	0.15
OHC 6	0.030	140.00	0.00	0	0.00	1985	0.15
OHC 8	0.030	170.00	0.00	0	0.00	1995	0.15
4C/4V	0.060	240.00	0.00	30	0.00	1990	0.30
6C/4V	0.060	320.00	0.00	45	0.00	1990	0.30
8C/4V	0.060	400.00	0.00	60	0.00	2002	0.30
Cylinder Reduction	0.030	-100.00	0.00	-150	0.00	1990	-0.10
4C/5V	0.080	300.00	0.00	45	0.00	1997	0.55
Turbo	0.050	500.00	0.00	80	0.00	1980	0.45
Engine Friction Reduction I	0.020	20.00	0.00	0	0.00	1991	0
Engine Friction Reduction II	0.035	50.00	0.00	0	0.00	2002	0
Engine Friction Reduction III	0.050	90.00	0.00	0	0.00	2012	0
Engine Friction Reduction IV	0.065	140.00	0.00	0	0.00	2022	0
VVT I	0.080	140.00	0.00	40	0.00	2006	0.10
VVT II	0.100	180.00	0.00	40	0.00	2016	0.15
Lean Burn	0.100	150.00	0.00	0	0.00	2099	0
Two Stroke	0.150	150.00	0.00	-150	0.00	2099	0
TBI	0.020	40.00	0.00	0	0.00	1985	0.05
MPI	0.035	80.00	0.00	0	0.00	1985	0.10
Air Pump	0.010	0.00	0.00	-10	0.00	1985	0
DFS	0.015	15.00	0.00	0	0.00	1985	0.10
Oil %w-30	0.005	2.00	0.00	0	0.00	1987	0
Oil Synthetic	0.015	5.00	0.00	0	0.00	1997	0
Tires I	0.010	16.00	0.00	0	0.00	1992	0
Tires II	0.020	32.00	0.00	0	0.00	2002	0
Tires III	0.030	48.00	0.00	0	0.00	2012	0
Tires IV	0.040	64.00	0.00	0	0.00	2018	0
ACC I	0.005	15.00	0.00	0	0.00	1997	0
ACC II	0.010	30.00	0.00	0	0.00	2007	0
EPS	0.015	40.00	0.00	0	0.00	2002	0
4WD Improvements	0.030	100.00	0.00	0	-0.05	2002	0
Air Bags	-0.010	300.00	0.00	35	0.00	1992	0
Emissions Tier I	-0.010	150.00	0.00	10	0.00	1996	0
Emissions Tier II	-0.010	300.00	0.00	20	0.00	2004	0
ABS	-0.005	300.00	0.00	10	0.00	1990	0
Side Impact	-0.005	100.00	0.00	20	0.00	1996	0
Roof Crush	-0.003	100.00	0.00	5	0.00	2001	0
Increased Size/Wt.	-0.200	0.00	0.00	0	0.30	1991	0
GDI/4-cyl	0.170	1000.00	0.00	0	0.00	2005	0
GDI/6-cyl	0.170	1200.00	0.00	0	0.00	2005	0
Gasoline Hybrid	0.450	0.00	75.00	0	0.05	2001	0

N/A = Non Applicable

[1]Fractional changes refer to the percentage change from the 1990 values.

Source: Energy and Environment Analysis, *Changes to the Fuel Economy Module*, Final Report, prepared for the Energy Information Administration (EIA), (June 1998).

The discounted stream of fuel savings is compared to the marginal cost of each technology. The fuel economy module assumes the following:

* All fuel saving technologies have a 4-year payback period.

* The real discount rate remains steady at 8 percent.

* Corporate Average Fuel Efficiency standards remain constant at 1998 levels.

* Expected future fuel prices are calculated based on an extrapolation of the growth rate between fuel prices 3 years and 5 years prior to the present year. This assumption is founded upon an assumed lead time of 3 to 5 years to significantly modify the vehicles offered by a manufacturer.

Degradation factors (Table 22) used to convert Environmental Protection Agency-rated fuel economy to actual "on the road" fuel economy are based on application of a logistic curve to the projections of three factors: increases in city/highway driving, increasing congestion levels, and rising highway speeds.[31] Degradation factors are also adjusted to reflect the percentage of reformulated gasoline consumed.

Table 22. Car and Light Truck Degradation Factors

	1998	2000	2005	2010	2015	2020
Cars	0.861	0.855	0.885	0.849	0.843	0.838
Light Trucks	0.814	0.807	0.808	0.804	0.797	0.791

Source: Green, Tamara, "Re-estimation of Annual Energy Outlook 2000 Degradation Factors," prepared for the Energy Information Administration, unpublished paper, August 18, 1999, Washington, D.C.

* The vehicle miles traveled (VMT) module forecasts VMT as a function of the cost of driving per mile, income per capita, ratio of female to male VMT, and age distribution of the driving population (Figure 4). Coefficients were re-estimated for *AEO2000* to adjust for the change in the definition of income to include stock equity and chain-weighting of income. The ratio of female to male VMT is assumed to asymptotically approach 80 percent by 2010. VMT per driver by age group was also assumed to be more uniformly distributed to older age groups. Total VMT is calibrated to Federal Highway Administration VMT data.[32,33] The fuel price elasticity rises from -0.05 to -0.2 as fuel prices rise above reference case levels in each year.

* The share of light truck sales is assumed to reach a maximum of 51 percent of total sales by 2020. However, the light truck share will gradually decline to 46 percent if fuel prices rise to approximately $1.50/gal. The size class sales shares will also gravitate to 25 percent for subcompacts, 40 percent for compacts, 25 percent for mid size, and 10 percent for luxury if fuel prices exceed reference case levels approximately $1.50/gal.

Commercial Light-Duty Fleet Assumptions

With the current focus of transportation legislation on commercial fleets and their composition, the Transportation Demand Module has been redesigned to divide commercial light-duty fleets into three types of fleets: business, government, and utility. Based on this classification, commercial light-duty fleet vehicles vary in survival rates and duration in the fleet, before being combined with the personal vehicle stock (Table 23). Sales shares of fleet vehicles by fleet type also remain constant over the forecast period. Automobile fleets are divided into the following shares: business (87.39%), government (7.42%), and utilities (5.19%). Light truck fleets are divided into the following shares: business (83.50%), government (14.1%), and utilities(2.40%)[34,35]. Both car (23.70%) and light truck (28.57%) fleet sales are assumed to be a constant fraction of total car and light truck sales.

Alternative-fuel shares of fleet sales by fleet type are initially set according to historical shares (business (0.36%), government (2.21%), utility (2.64%))[36,37] then compared to a minimum constraint level of sales

Table 23. The Average Length of Time Vehicles Are Kept Before they are Sold to Others
(Months)

Vehicle Type	Business	Utility	Government
Cars	35	68	81
Light Trucks	56	60	82
Medium Trucks	83	86	96
Heavy Trucks	103	132	117

Source: Oak Ridge National Laboratory, *Fleet Vehicles in the United States: Composition, Operating Characteristics, and Fueling Practices*, prepared for the Department of Energy, Office of Transportation Technologies and Office of Policy, Planning, and Analysis (Oak Ridge, TN, May 1992).

based on legislative initiatives, such as the Energy Policy Act and the Low Emission Vehicle Program.[38],[39] Size class sales shares of vehicles are held constant at anticipated levels (Table 24).[40],[41] Individual sales shares of alternative-fuel fleet vehicles by technology type are assumed to remain at anticipated levels for utility, government, and for business fleets in accordance with the technology shares implied from EIA surveys[42],[43] (Table 25).

Annual VMT per vehicle by fleet type stays constant over the forecast period based on the Oak Ridge National Laboratory fleet data.

Table 24. Commercial Fleet Size Class Shares by Fleet and Vehicle Type, 1992
(Percentage)

Fleet Type by Size Class	Automobiles	Light Trucks
Business Fleet		
Small	4.55	37.34
Medium	71.59	37.90
Large	23.86	24.76
Government Fleet		
Small	4.35	21.34
Medium	56.52	44.39
Large	39.13	34.27
Utility Fleet		
Small	16.67	30.03
Medium	70.00	38.51
Large	13.33	31.46

Source: Oak Ridge National Laboratory, *Fleet Vehicles in the United States: Composition, Operating Characteristics, and Fueling Practices*, unpublished final report prepared for the Department of Energy, Office of Transportation Technologies and Office of Policy, Planning, and Analysis, (Oak Ridge, TN, May 1992).

Table 25. Anticipated Purchases of Alternative-Fuel Vehicles by Fleet Type and Technology Type
(Percentage)

AFV Technology	Business	Government	Utility
Ethanol	0.02	3.06	0.00
Methanol	1.62	21.98	3.37
Electric	0.90	0.19	3.10
CNG	9.46	58.73	66.94
LPG	88.00	16.04	26.58

Sources: Energy Information Administration, *Describing Current and Potential Markets for Alternative Fuel Vehicles*, DOE/EIA-0604(96), (Washington, DC, March 1996). Energy Information Administration, *Alternatives to Traditional Transportation Fuels 1996*, DOE/EIA-0585(96), (Washington, DC, December 1997).

Fleet fuel economy for both conventional and alternative-fuel vehicles is assumed to be the same as the personal new vehicle fuel economy and is subdivided into six EPA size classes for car and light truck.

Figure 4. VMT per Driver by Age-Group
     (Vehicles-Miles Traveled)

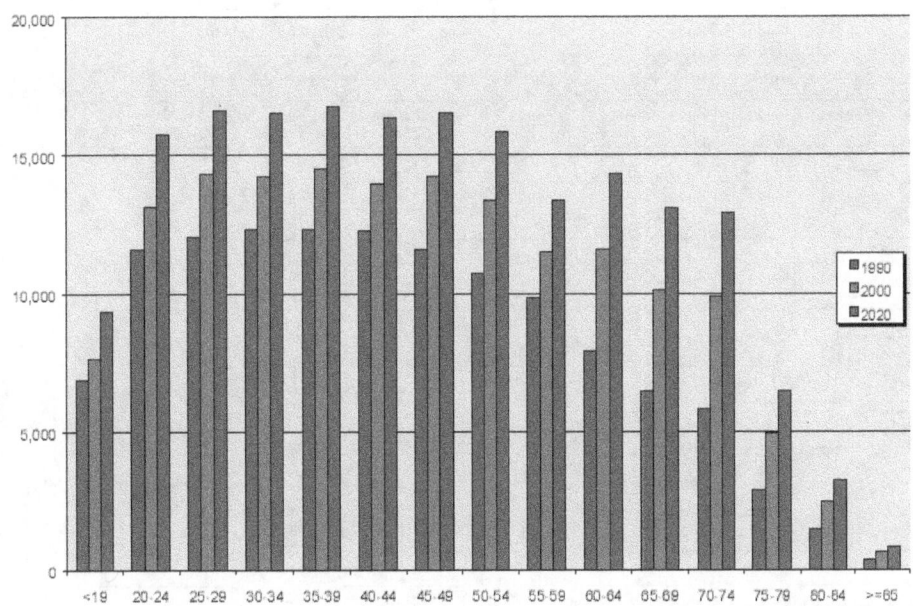

Source: 1990 values: U.S. Dept. of Transportation, Summary of Travel Trends: *1995 National Personal Transportation Survey*, draft, prepared by Oak Ridge National Laboratory Washington D.C. 1999; Forecast: EIA, *AEO2000* National Energy Modeling System run: aeo2k.d100199a.

The Light Commercial Truck Module

The Light Commercial Truck Module of the NEMS Transportation Model is constructed to represent trucks that weight 8501 lbs. to 10,000 lbs. These vehicles are assumed to be used for commercial freight purposes.

The primary source of data for this model is the microdata file of the 1992 Truck Inventory and Use Survey (TIUS), which provides numerous details on truck stock and usage patterns at a high level of disaggregation. The data derived from this source are used to allocate and sort the summary truck data presented in the Federal Highway Administration's annual publication of highway statistics, which constitute the baseline from which the NEMS forecast is made (Figure 5). TIUS data are also used to distribute estimated sales of trucks, obtained from the Macroeconomic Model, among the affected models according to their weight class (Figure 6). Finally, the TIUS microdata set is used to construct a characterization of these Light Commercial Trucks.

Truck characterizations comprised of their average annual miles of travel, fuel economy, and distribution among several aggregate industrial groupings chosen for their correspondence with output measures currently being forecast by NEMS (Tables 26 and 27). It is expected that projected growth in industrial output will provide a useful proxy for the growth in demand for the services of light commercial trucks.

VMT for light commercial trucks is a function of industrial output for agriculture, mining, construction, trade, utilities, and personal VMT. Forecasted fuel efficiencies are assumed to increase at the same annual growth rate as light-duty trucks (<8500 lbs.).

Figure 5. Distribution of FHWA Single-Unit Truck Stocks

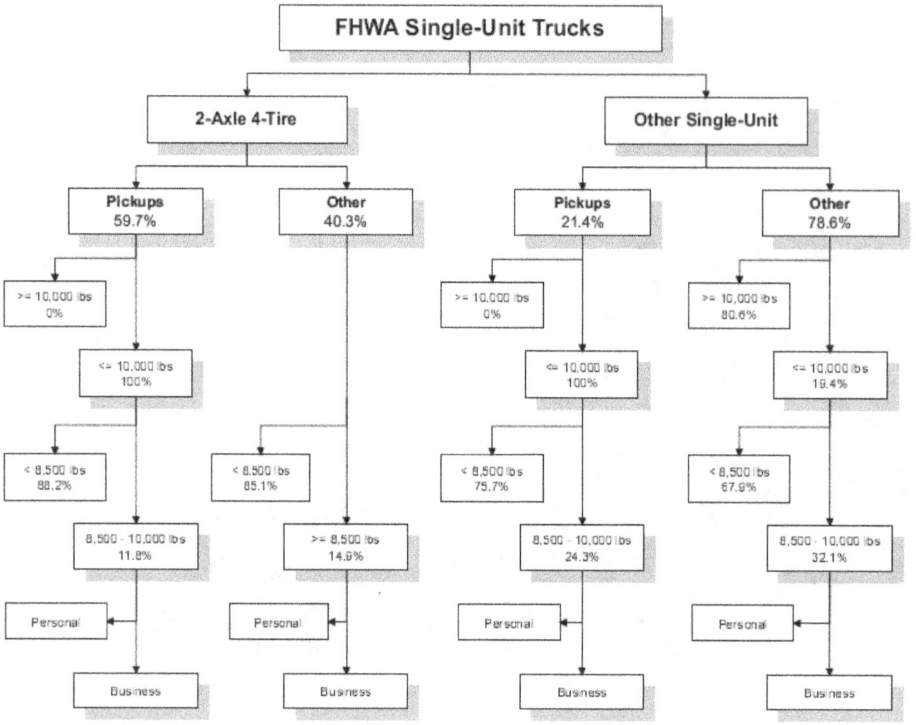

FHWA Single-Unit Trucks
2-Axle 4-Tire
Other Single-Unit
Pickups 59.7%
Other 40.3%
Pickups 21.4%
Other 78.6%
>= 10,000 lbs 0%
<= 10,000 lbs 100%
< 8,500 lbs 88.2%
8,500 - 10,000 lbs 11.8%
< 8,500 lbs 85.1%
>= 8,500 lbs 14.9%
>= 10,000 lbs 0%
<= 10,000 lbs 100%
< 8,500 lbs 75.7%
8,500 - 10,000 lbs 24.3%
>= 10,000 lbs 80.6%
<= 10,000 lbs 19.4%
< 8,500 lbs 67.9%
8,500 - 10,000 lbs 32.1%
Personal
Business
Personal
Business
Personal
Business
Personal
Business

Source: U.S. Dept. Of Transportation, Federal Highway Administration, Highway Statistics 1995, Nov. 1996; U.S. Dept. Of Commerce, Bureau of the Census, Truck Inventory and Use Survey 1992,

Figure 6. Distribution of Light Truck Sales

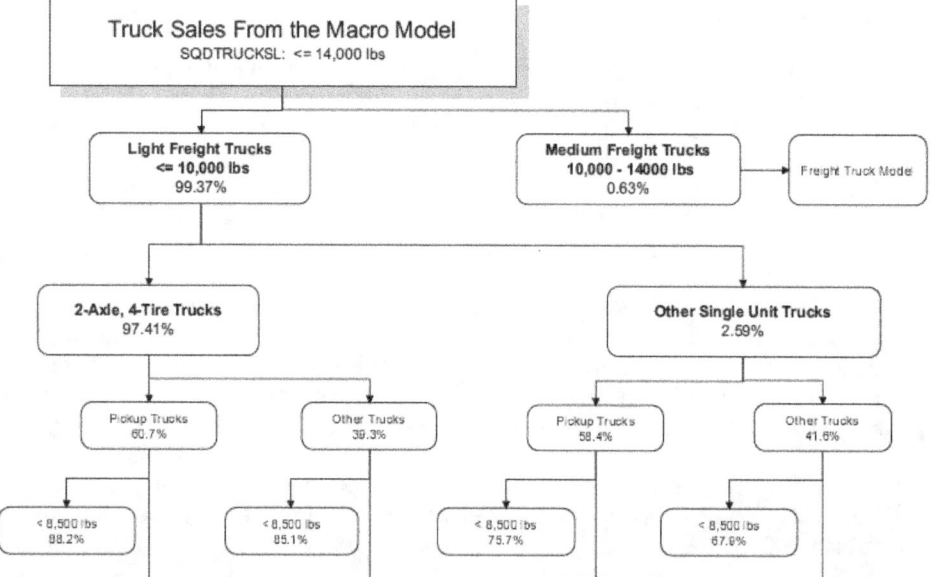

Truck Sales From the Macro Model
SQDTRUCKSL: <= 14,000 lbs
Light Freight Trucks <= 10,000 lbs 99.37%
Medium Freight Trucks 10,000 - 14000 lbs 0.63%
Freight Truck Model
2-Axle, 4-Tire Trucks 97.41%
Other Single Unit Trucks 2.59%
Pickup Trucks 60.7%
Other Trucks 39.3%
Pickup Trucks 58.4%
Other Trucks 41.6%
< 8,500 lbs 88.2%
< 8,500 lbs 85.1%
< 8,500 lbs 75.7%
< 8,500 lbs 67.9%
8,500 - 10,000 lbs 11.8%
>= 8,500 lbs 14.9%
8,500 - 10,000 lbs 24.3%
8,500 - 10,000 lbs 32.1%

Source: U.S. Dept. Of Transportation, Federal Highway Administration, Highway Statistics 1995, Nov. 1996; U.S. Dept. Of Commerce, Bureau of the Census, Truck Inventory and Use Survey 1992,

Table 26. Anticipated Annual Miles, by Major Use (1992 TIUS)
         (Aggregated for NEMS)

| Major Use | Single-Unit Trucks, 6,000 - 10,000 lbs. | | | |
| | 2 Axle, 4 Tire | | Other Single-Unit | |
	Pickup	Other	Pickup	Other
Agriculture	11,920	8,569	15,197	7,054
Mining	20,231	24,871	18,520	17,786
Construction	15,909	15,195	13,043	10,074
Trade	13,313	15,394	10,009	11,832
Utilities	13,023	13,776	9,947	9,996
Personal	9,980	10,148	8,429	5,852

Source: 1992 TIUS- U.S. Dept. Of Transportation, Federal Highway Administration, Highway Statistics 1995, Nov. 1996; U.S. Dept. Of Commerce, Bureau of the Census, Truck Inventory and Use Survey 1992, TC-92-T-52, (Washington DC., May 1995).

Table 27. Average Miles Per Gallon Estimate

| Major Use | 2 Axle, 4 Tire | | | |
	Pickup	Other	Pickup	Other
Agriculture	12.77	8.75	11.79	8.66
Mining	13.12	11.92	12.00	10.10
Construction	13.45	11.79	12.58	8.92
Trade	13.55	11.57	12.71	8.98
Utilities	13.33	10.25	13.57	8.65
Personal	13.67	13.99	12.29	10.78

Source: U.S. Dept. Of Commerce, Bureau of the Census, Truck Inventory and Use Survey 1992, TC-92-T-52, (Washington, DC., May 1995).

Alternative-Fuel Vehicle Technology Choice Assumptions

Fuel vehicle (AFV) technology choice module utilizes a discrete choice specification, which uses vehicle attributes as inputs and forecasts vehicle sales shares among the following 15 light-duty technologies: gasoline internal combustion engine (ICE), direct injection diesel ICE, ethanol flex, ethanol neat, methanol flex, methanol neat, electric dedicated (uses only electricity), diesel electric hybrid, compressed natural gas (CNG), CNG bi-fuel, LPG, LPG bi-fuel, fuel cell gasoline, fuel cell methanol, and fuel cell liquid hydrogen.[44] Direct injection gasoline and gasoline electric hybrid technologies are included in the conventional gasoline ICE technologies.

Listed in Table 28 are a few examples of the input variables that correspond to the vehicle attributes used in the analysis. With the exception of vehicle fuel economy, fuel price, vehicle price, vehicle range, 0-30 second acceleration times, and top speed, all other attributes are exogenously set, based on offline analysis.[45]

Vehicle attributes vary by six EPA size classes for cars and light trucks, and fuel availability varies by Census Division. The logit model coefficients vary by three car sizes and four light truck sizes. Vehicle prices are assumed to follow exponential curves of economies of scale in production dependent upon the volumes and cost curves which vary by AFV technologies. Where applicable, AFV fuel efficient technology attributes are calculated relative to conventional gasoline miles per gallon. It is assumed that many fuel efficiency improvements to conventional vehicles will be transferred to alternative-fuel vehicles. Specific individual alternative-fuel technological improvements are also dependent upon the AFV technology type, cost, research and development, and availability over time. Commercial availability estimates are assumed values according to a logistic curve based on the initial technology introduction date and were constructed in cooperation with the Office of Energy Efficiency and Renewable Energy of the Department of Energy (DOE). Coefficients summarizing consumer valuation of vehicle attributes were derived from a stated preference survey conducted in the U.S.[46] and are assumed to be representative of the United States. Initial AFV vehicle stocks are set according to EIA surveys.[47,48] A fuel switching algorithm based on the relative fuel prices for AF compared to gasoline is used to detrmine the percentage of total VMT represented by AFin bi-fuel and flex-fuel alcohol vehicles. An upper limit of 50 percent and a lower limit of 25 percent is assumed for the percentage of the vehicle-miles traveled using the alternative fuel.

Table 28 . Alternative-Fuel Vehicle Attribute Inputs For Two Stage Logit Model

Vehicle Attributes	Year	Gasoline	Ethanol Flex	Methanol Flex	CNG	Diesel Electric Vehicle Hybrid	Fuel Cell Gasoline
Vehicle Price (thousand 1990 dollars)	1998	16.40	16.80	16.80	20.20	N/A	N/A
	2020	18.00	18.40	18.40	21.80	29.70[1]	33.00[1]
Vehicle MPG (miles/gallon)	1998	30.66	29.77	30.07	31.69	N/A	N/A
	2020	33.39	31.83	32.12	33.45	51.60	47.02
Vehicle Range (100 miles)	1998	4.60	3.24	2.53	2.66	N/A	N/A
	2020	5.01	3.47	2.71	2.85	6.18	4.75
Fuel Availability Relative to Gasoline	1998	1.00	1.00	1.00	0.02	1.00	1.00
	2020	1.00	1.00	1.00	0.22	1.00	1.00
Commercial Availability Indexed To Gasoline	1998	1.00	0.007	0.007	0.001	N/A	N/A
	2020	1.00	0.999	0.999	0.993	0.999	0.999

[1]Electric vehicle battery replacement cost included.

CNG = Compressed natural gas.

MPG = Miles per gallon.

N/A = Not Available Commercially.

Sources: Vehicle prices, fuel efficiency, and range: Energy and Environmental Analysis, Updates to the *Fuel Economy Module Final Report*, Prepared for EIA, (June 1998).

Freight Truck Assumptions

The freight stock truck module converts industrial output in dollar terms to an equivalent measure of volume by using a freight adjustment coefficient.[49,50]These freight truck adjustment coefficients vary by NEMS Standard Industrial Classification (SIC) code, gradually diminishing their deviation over time toward parity and are estimated from historical freight data. Freight truck load factors (ton-miles per truck) by SIC code are constants formulated from historical load factors.[51] All freight trucks are subdivided into medium and heavy-duty trucks. New freight truck fuel efficiency is dependent on the maximum penetration, introduction year, cost-effectiveness based on fuel price and capital costs, and fuel economy improvement of the technologies including alternative fuel technologies (Table 29).[52] VMT freight estimates by size class and technology are based on matching freight needs as measured by the growth in industrial output by SIC code to VMT levels associated with truck stocks and new vehicles. Fuel consumption by freight trucks is regionalized by Census Division according to the *State Energy Data Report* distillate regional shares.[53]

Initial freight trucks are obtained by the Federal Highway Administration (FHWA) and are distributed by Truck and Inventory Use Survey (TIUS) shares.

Table 29. Diesel Technology Characteristics for the Freight Truck Model

	Fuel Economy Improvement (%)		Maximum Penetration (%)		Introduction Yr		Capital Cost (1988 dollars)	
	Medium	Large	Medium	Large	Medium	Large	Medium	Large
Existing Technologies								
Advanced Tires: Radials	2	2	70	70			$150	$450
Drag Reduction	3	5	65	65			$500	$1,000
New Technologies								
Advanced Transmissions	2	1	40	40	2001	2001	$2,500	$2,500
Lightweight Materials	1	1	30	30	2002	2002	$3,000	$3,000
Synthetic Gear Lube	2	2	60	60	2001	2001	$40	$60
Advanced Tires: Low Resistance	4	4	70	70	2001	2001	$300	$900
Advanced Drag Reduction	4	7	65	65	2001	2002	$ 600	$1,200
Electronic Engine Control	4	4	95	95	2001	2001	$1,000	$1,000
Advanced Engine	9	9	90	90	2009	2009	$1,000	$1,000
Turbocompounding	0	5	0	90	N/A	2001	N/A	$2,000
Hybrid Powertrain	54	0	20	0	2006	N/A	$6,000	N/A
Port-Injection	0	1	0	100	N/A	2001	N/A	$300

Source: Argonne National Laboratories, Frank Stodolsky, Anant Vyas, Roy Cuenca. *"Heavy and Medium Duty Truck Fuel economy and Market Penetration Analysis"*, prepared for Energy Information Administration, August 6, 1999.

Freight and Transit Rail Assumptions

The freight rail module receives industrial output by SIC code measured in real 1987 dollars and converts these dollars into an adjusted volume equivalent. Specific NEMS coal production from the Coal Module is also used to adjust coal rail travel. Freight rail adjustment coefficients, which are used to convert dollars into volume equivalents, remain constant and are based on historical data.[54],[55] Initial freight rail efficiencies are based on the freight model from Argonne National Laboratory.[56] The distribution of rail fuel consumption by fuel type remains constant and is based on historical data.[57] Regional freight rail consumption estimates are distributed according to the *State Energy Data Report 1994*.[58]

Freight Domestic and International Shipping Assumptions

The freight domestic shipping module also converts industrial output by SIC code measured in dollars, to a volumetric equivalent by SIC code.[59],[60] These freight adjustment coefficients are based on analysis of historical data and remain constant throughout the forecast period. Domestic shipping efficiencies are based on the freight model by Argonne National Laboratory. The energy consumption in the freight international shipping module is a function of the total level of imports and exports. The distribution of domestic and international shipping fuel consumption by fuel type remains constant throughout the analysis and is based on historical data.[61] Regional domestic and international shipping consumption estimates are distributed according to the *State Energy Data Report* residual oil regional shares.[62]

Air Travel Demand Assumptions

The air travel demand module calculates the ticket price for travel as a function of fuel cost. Similar to the light-duty vehicle module, the air travel fuel price elasticity rises from -0.04 to -0.2 if jet fuel prices exceed reference case levels. A demographic index based on the propensity to fly was introduced into the air travel demand equation.[63] The propensity to fly was made a function of the age and gender distribution over the forecast period[64],[65] The air travel demand module assumes that these relationships between the groups and their propensity to fly remain constant over time. International revenue passenger miles are calculated as a percentage of domestic revenue passenger miles based on an extrapolation of historical data, which asymptotically approaches 56 percent by 2020.[66] The revenue ton miles of air freight are based on merchandise exports and gross domestic product.

Aircraft Stock/Efficiency Assumptions

The aircraft stock and efficiency module consists of a stock model of both wide and narrow body planes by vintage. The shifting of passenger load between narrow and wide body aircraft is assumed to occur at a constant historical annual 1-percent rate.[67] The available seat-miles per plane, which measure the carrying capacity of the airplanes by aircraft type, remain constant and are based on holding the seat-miles and the number of planes constant within an aircraft type.[68] The difference between the seat-miles demanded and the available seat-miles represents newly purchased aircraft. Aircraft purchases in a given year cannot exceed historical annual growth rates, a constraint that sets an upper limit on the application of new aircraft to meet the gap between seat-miles demanded and available seat-miles. With a constraint on new aircraft purchases, it is assumed that when the gap exceeds historical aircraft sales levels, planes that have been temporarily stored or retired will be brought back into service. Technological availability, economic viability, and efficiency characteristics of new aircraft are based on the technologies listed in the Oak Ridge National Laboratory Air Transport Energy Use Model. (See Table 33 on page 59)[69] Fuel efficiency of new aircraft acquisitions represents, at a minimum, a 5-percent improvement over the stock efficiency of surviving airplanes.[70] Maximum growth rates of fuel efficiency for new aircraft are based on a future technology improvement list consisting of an estimate of the introduction year, jet fuel price, and an estimate of the proposed marginal fuel efficiency improvement. Regional shares of all types of aircraft fuel are assumed to be constant and are consistent with the *State Energy Data Report* estimate of regional jet fuel shares.

Legislation

Energy Policy Act of 1992 (EPACT)

Fleet alternative-fuel vehicle sales necessary to meet the EPACT regulations were derived based on the mandates as they currently stand and the Commercial Fleet Vehicle Module calculations. Total projected AFV sales are divided into fleets by government, business, and fuel providers (Table 30). Business fleet EPACT mandates are not included in the projections for AFV sales pending a decision on a proposed rulemaking.

Table 30. EPACT Legislative Mandates for Percentage AFV Purchases by Fleet Type, Year

Year	Municipal & Business	Federal	State	Fuel Providers	Electric Utilities
1996	-	25	-	-	-
1997	-	33	10	30	-
1998	-	50	15	50	30
1999	-	75	25	70	50
2000	-	75	50	90	70
2001	-	75	75	90	90
2002	20	75	75	90	90
2003	40	75	75	90	90
2004	60	75	75	90	90
2005	70	75	75	70	90

Source: EIA, *Alternatives to Traditional Transportation Fuels 1994*, DOE/EIA-0585(94), (Washington, D.C, February 1996).

Because the commercial fleet model operates on three fleet type representations (business, government, and utility), the federal and state mandates were weighted by fleet vehicle stocks to create a composite mandate for both. The same combining methodology was used to create a composite mandate for electric utilities and fuel providers based on fleet vehicle stocks.[71],[72] Fleet vehicle stocks by car and light truck were disaggregated to include only fleets of 50 or more (in accordance with EPACT) by using a fleet size distribution function based on The Fleet Factbook and the Truck and Inventory Use Survey.[73],[74] To account for the EPACT regulations which stipulate that "covered" fleets (which refer to fleets bound by the EPACT mandates) include only fleets in the metropolitan statistical areas (MSA's) of 250,000 population or greater, 90 percent of the business and utility fleets were included and 63 percent were included for government fleets.[75] EPACT covered fleets were to only include those fleets that could be centrally fueled, which was assumed to be 50 percent of the fleets for all fleet types, and only fleets of 50 or more that had 20 vehicles or more in those MSA's of 250,000 or greater population; it was assumed that 90 percent of all fleets were within this category except for business fleets, which were assumed to be 75 percent.[76]

Low Emission Vehicle Program (LEVP)

The LEVP, which began in California, which was originally instituted in New York and Massachusetts, has now been rolled back to begin in 2003 at the original 10 percent mandate for California, Massachusetts and New York. The following Zero Emission Vehicle (ZEV) sales percentage numbers (Table 31) come from the California Air Resources Board.[77] All of the ULEV sales were assumed to meet the ULEV air standards with reformulated gasoline and a heated catalytic converter.

Table 31. Original and Revised California Low Emission Vehicle Program Legislatively Mandated
Alternative-Fuel Vehicle Sales
(Percentage of all sales)

Vehicle	1997	1998	1999	2000	2001	2002	2003
Original							
Zero Emission Vehicles	--	2	2	2	5	5	10
Revised							
Zero Emission Vehicles	--	--	--	--	--	--	10

Source: California Air Resources Board, *Proposed Regulations for Low Emission Vehicles and Clean Fuels, Staff Report*, August 13, 1990.

On November 5, 1998, the California Air Resources Board (CARB) amended the original LEVP to include ZEV credits for advanced technology vehicles. According to CARB these advanced technology vehicles must be capable of achieving "extremely low levels of emissions on the order of the power plant emissions that occur from charging battery-powered electric vehicles, and some that demonstrate other ZEV-like characteristics such as inherent durability and partial zero-emission range."[78]

There are three components to calculating the ZEV credit, a baseline ZEV allowance, a zero-emission vehicle-miles traveled (VMT) allowance, and a low fuel-cycle emission allowance. Using these advanced vehicles in place of ZEV's in order to comply with the LEVP mandates requires assessment of each vehicle characteristic relative to the three criteria allowances.

The baseline ZEV allowance potentially can provide up to .2 credits if the advanced technology vehicle meets the: a) Super Ultra Low Emission Vehicle (SULEV) standards contained in the originial LEVP proposal; b) on-board diagnostics requirements (OBD) which illuminates indicators on the dashboard when vehicles are out of emissions compliance levels; c) 150,000 mile emission equipment warranty; and d) evaporative emissions requirements in California which prevent emissions during refueling. SULEV emissions standards approximate the emissions from powerplants associated with recharing electric vehicles.

The second criteria, zero-emission VMT allowance, will allow a maximum .6 credit if the vehicle is capable of some all-electric operation which was fueled by off-vehicle sources (i.e. no on-board fuel reformers), or if the vehicle has ZEV-like equipment on-board such as regenerative braking, advanced batteries, or an advanced electric drivetrain.

An emission allowance was also made for low fuel-cycle vehicle fuels used in the advanced technology vehicles. A maximum of .2 credit is provided for vehicles which use fuel that has less than or equal to .01 NMOG grams per mile emissions based on the grams per gallon and the fuel efficiency of the vehicle.

Overall, large volume manufacturers can apply ZEV credits up to a maximum of 60 percent of the original 10 percent ZEV mandate; the original ZEV mandate required that all (100 percent) of the 10 percent of all light-duty vehicle sales must be ZEVs (defined only as dedicated electric vehicles) beginning with the 2003 model year. The remaining 40 percent of the ZEV mandates must still come from electric vehicles, or variants of cell vehicles, which have extremely low emissions such as a hydrogen fuel cell vehicle.

The AFV sales module compares these legislatively mandated sales to the results from the AFV logit market-driven sales shares. The legislatively mandated sales serve as a minimum constraint to AFV sales.

According to the EPA federal register, EPA's Tier II proposed regulations for light-duty vehicles below 6000 pounds must meet a sales weighted average of 0.07 grams/mile NOx emissions standard by 2004 and approximately a 0.01 to 0.02 grams/mile standard for particulates.[79] The previous Clean Air Act 1990 Tier I emissions standards were set at 0.6 grams/mile for NOx and 0.1 grams/mile for particulates.[80] EPA has estimated the costs to consumers range from $100 per car to $200 per light-truck.[81] However, recently the U.S. Circuit Court ruling determined that EPA was not authorized to set new standards without indicating the benefits of the new regulations.

In the National Research Council's (NRC) Fifth Annual Review of Partnership for a New Generation of Vehicles (PNGV)[82], the NRC committee commented,"..the most difficult technical challenge facing the CIDI (compression ignition direct injection diesel) engine program will be meeting the standards for NOx and particulate emissions. In addition, meeting an even more stringent research objective (0.01 grams/mile) for particulate matter instead of the 0.04 grams/mile PNGV target would require additional technological breakthroughs."

The NRC has stated their concern that the Tier II regulations may affect the commercial viability of many advanced vehicles. Meeting the Tier II proposed standards may: require trading-off emissions levels for fuel economy by redesigning engines; add significant cost to a technology due to exhaust catalyst systems and their potential lack of effectiveness; stifle development of diesel technologies as a result of the unknown health effects of particulates; and result in new specifications for diesel fuel or development of advanced low emission fuels.

Climate Change Action Plan

There were four programs implemented from the Climate Change Action Plan (CCAP) transportation policies—reform Federal subsidy for employer-provided parking, adopt a transportation system efficiency strategy, promote telecommuting, and develop fuel economy labels for tires. The combined effect of the Federal subsidy, system efficiency, and telecommuting policies was a reduction in VMT of 1.6 percent in 2010, representing a decline in consumption of approximately 140 trillion Btu with a net carbon reduction of 2.8 million metric tons. The fuel economy tire labeling program improved fuel efficiency by 4 percent among vehicles that switched to low rolling resistance tires in pre-1999 vehicles. Therefore there are no new fuel or carbon savings from this program.

Advanced Technology and 2000 Technology Cases

In the *advanced technology case*, the light-duty vehicle assumptions for alternative fuel vehicles are presented in Table 32 and are based on the yearly U.S. Department of Energy Office of Energy Efficiency and Renewables Office of Transportation Technologies (OTT) Program Analysis[83] The conventional fuel saving technology characteristics come from a study by the American Council For an Energy Efficient Economy.[84] In the *advanced technology case,* fuel efficiency improvements from new technology more than offset the increasing travel in each transportation mode. As a result, the total energy consumption in the transportation sector was 11.1 percent lower (4.15 quadrillion Btu) than in the reference case by 2020. Tables 34 and 35 summarize the High Technology matrix for cars and trucks.

The *2000 technology case* assumes that new fuel efficiency technologies are held constant at 2000 levels over the forecast. As a result, the energy use in the transportation sector was 6.0 percent higher (2.27 quadrillion Btu) than in the reference case by 2020. Both cases were run with only the transportation demand module rather than as a fully integrated NEMS run. Consequently, no potential macroeconomic feedback on travel demand, or fuel economy was captured.

Freight trucks in the *advanced technology case* were constructed in accordance with the assumptions from a Department of Energy (DOE) study.[85] The following technologies were made commercially available within the forecast period: advanced drag reduction, turbocompound diesel engine, heat engine CLE-55,

Table 32. Advanced Technology Alternative-Fuel Large Car Vehicle Assumptions Relative to Conventional Gasoline Vehicle, 2020
(Thousands)

Technology	Year of Introduction	Year of Maturity	Vehicle Cost Ratio	Fuel Economy Ratio	Relative Vehicle Range
Advanced Diesel	2004	2017	Intro:1.11 Mat.:1.04	Intro: 1.35 Mat.: 1.35	Intro: 1.14 Mat.: 1.20
Diesel Hybrid	2003	2020	Intro: 1.40 Mat.: 1.16	Intro: 1.5 Mat.: 2.00	Intro: 1.20 Mat.: 1.20
Fuel Cell	2007	2019	Intro: 1.50 Mat.: 1.26	Intro: 2.10 Mat.: 2.20	Intro: 1.00 Mat.: 1.00
Natural Gas	2000	2006	Intro: 1.11 Mat.: 1.03	Intro: 1.0 Mat.: 1.0	Intro: 0.66 Mat.: 0.75
Flex Alcohol	1998	1998	Intro: 1.0 Mat.: 1.0	Intro: 1.0 Mat.: 1.0	Intro: 1.0 Mat.: 1.0

Source: U.S. Department of Energy, Office of Energy Efficiency and Renewables, Office of Transportation Technologies, *OTT Program Analysis Methodology: Quality Metrics 2000*, November 1, 1998.

and reduced empty weight technologies. Additionally, shorter market penetration periods, and technology prices were made cost-effective at $6/MMBtu for diesel fuel, instead of the range of $8-28.60/MMBtu in the *AEO99* reference case.

The air model in the *advanced technology case* assumed efficiency from new aircraft could improve by 40 percent from the 1992 level based on the conclusion from the Aeronautics and Space Engineering Board of the National Research Council.[86]

Table 33. Future New Aircraft Technology Improvement List

Proposed Technology	Introduction Year	Jet Fuel Price Necessary For Cost-Effectiveness (1987 dollars per gallon)	Seat-Miles per Gallon Gain Over 1990 (percent)	
			Narrow Body	Wide Body
Engines				
Ultra-high Bypass	1995	$.69	10	10
Propfan	2000	$1.36	23	0
Thermodynamics	2010	$1.22	20	20
Aerodynamics				
Hybrid Laminar Flow	2020	$1.53	15	15
Advanced Aerodynamics	2000	$1.70	18	18
Other				
Weight Reducing Materials	2000	-	15	15

Source: Greene, D.L., *Energy Efficiency Improvement Potential of Commercial Aircraft to 2010*, ORNL-6622, 6/1990., and from data tables in the Air Transportation Energy Use Model (ATEM), Oak Ridge National Laboratory.

Table 34. High Technology Matrix For Trucks

	Fractional Fuel Efficiency Change	Incremental Cost (1990 $)	Incremental Cost ($/Unit Wt.)	Incremental Weight (Lbs.)	Incremental Weight (Lbs./Unit Wt.)	First Year Introduced	Fractional Horsepower Change
Front Wheel Drive	0.020	160.00	0.00	0	-0.08	1985	0
Unit Body	0.060	80.00	0.00	0	-0.05	1995	0
Material Substitution II	0.033	0.00	0.60	0	-0.05	1986	0
Material Substitution III	0.066	0.00	0.80	0	-0.10	2006	0
Material Substitution IV	0.099	0.00	1.00	0	-0.15	2016	0
Material Substitution V	0.132	0.00	1.50	0	-0.20	2026	0
Drag Reduction II	0.023	32.00	0.00	0	0.00	1990	0
Drag Reduction III	0.046	64.00	0.00	0	0.05	1997	0
Drag Reduction IV	0.069	112.00	0.00	0	0.01	2007	0
Drag Reduction V	0.092	176.00	0.00	0	0.02	2017	0
TCLU	0.030	40.00	0.00	0	0.00	1980	0
4-Speed Automatic	0.045	225.00	0.00	30	0.00	1980	0.05
5-Speed Automatic	0.065	325.00	0.00	40	0.00	1997	0.07
CVT	0.100	250.00	0.00	20	0.00	2005	0.07
6-Speed Manual	0.020	100.00	0.00	30	0.00	1997	0.05
Electronic Transmission I	0.005	20.00	0.00	5	0.00	1991	0
Electronic Transmission II	0.015	40.00	0.00	5	0.00	2006	0
Roller Cam	0.020	16.00	0.00	0	0.00	1986	0
OHC 4	0.030	100.00	0.00	0	0.00	1980	0.2
OHC 6	0.030	140.00	0.00	0	0.00	1985	0.2
OHC 8	0.030	170.00	0.00	0	0.00	1995	0.2
4C/4V	0.060	240.00	0.00	30	0.00	1990	0.45
6C/4V	0.060	320.00	0.00	45	0.00	1990	0.45
8C/4V	0.060	400.00	0.00	60	0.00	2002	0.45
Cylinder Reduction	0.030	-100.00	0.00	-150	0.00	1990	-0.1
4C/5V	0.080	300.00	0.00	45	0.00	1997	0.55
Turbo	0.050	500.00	0.00	80	0.00	1980	0.45
Engine Friction Reduction I	0.020	20.00	0.00	0	0.00	1991	0
Engine Friction Reduction II	0.035	50.00	0.00	0	0.00	2002	0
Engine Friction Reduction III	0.050	90.00	0.00	0	0.00	2012	0
Engine Friction Reduction IV	0.065	140.00	0.00	0	0.00	2022	0
VVT I	0.080	140.00	0.00	40	0.00	2006	0.1
VVT II	0.100	180.00	0.00	40	0.00	2016	0.15
Lean Burn	0.150	150.00	0.00	0	0.00	2018	0
Two Stroke	0.150	150.00	0.00	-150	0.00	2008	0
TBI	0.020	40.00	0.00	0	0.00	1985	0.05
MPI	0.035	80.00	0.00	0	0.00	1985	0.1
Air Pump	0.010	0.00	0.00	-10	0.00	1985	0
DFS	0.015	15.00	0.00	0	0.00	1985	0.1
Oil 5W-30	0.005	2.00	0.00	0	0.00	1987	0
Oil Synthetic	0.015	5.00	0.00	0	0.00	1997	0
Tires I	0.010	16.00	0.00	0	0.00	1992	0
Tires II	0.020	32.00	0.00	0	0.00	2002	0
Tires III	0.030	48.00	0.00	0	0.00	2012	0
Tires IV	0.040	64.00	0.00	0	0.00	2018	0
ACC I	0.005	15.00	0.00	0	0.00	1997	0
ACC II	0.010	30.00	0.00	0	0.00	2007	0
EPS	0.015	40.00	0.00	0	0.00	2002	0
4WD Improvements	0.030	100.00	0.00	0	-0.05	2002	0
Air Bags	-0.010	300.00	0.00	35	0.00	1992	0
Emissions Tier I	-0.010	150.00	0.00	10	0.00	1996	0
Emissions Tier II	-0.010	300.00	0.00	20	0.00	2004	0
ABS	-0.005	300.00	0.00	10	0.00	1990	0
Side Impact	-0.005	100.00	0.00	20	0.00	1996	0
Roof Crush	-0.003	100.00	0.00	5	0.00	2001	0
Increased Size/Wt.	-0.033	0.00	0.00	0	0.05	1991	0
GDI/4-cyl	0.170	1000.00	0.00	0	0.00	2005	0.02
GDI/6-cyl	0.170	1200.00	0.00	0	0.00	2005	0
Gasoline Hybrid	0.450	0.00	75.00	0	0.05	2001	0

Source: Energy and Enviromental Analysis, *Changes to the Fuel Economy Module, Final Report, 12-3, prepared for Energy Information Administration (EIA), (June 1998).*

Table 35. High Technology Matrix For Cars

	Fractional Fuel Efficiency Change	Incremental Cost (1990 $)	Incremental Cost/ ($/Unit Wt.)	Incremental Weight (Lbs.)	Incremental Weight (Lbs./ Unit Wt.)	First Year Introduced	Fractional Horsepower Change
Front Wheel Drive	0.060	160.00	0.00	0	-0.08	1980	0
Unit Body	0.040	80.00	0.00	0	-0.05	1980	0
Material Substitution II	0.033	0.00	0.60	0	-0.05	1987	0
Material Substitution III	0.066	0.00	0.80	0	-0.10	1997	0
Material Substitution IV	0.099	0.00	1.00	0	-0.15	2007	0
Material Substitution V	0.132	0.00	15.0	0	-0.20	2017	0
Drag Reduction II	0.023	32.00	0.00	0	0.00	1985	0
Drag Reduction III	0.046	64.00	0.00	0	0.05	1991	0
Drag Reduction IV	0.069	112.00	0.00	0	0.01	2004	0
Drag Reduction V	0.092	176.00	0.00	0	0.02	2014	0
TCLU	0.030	40.00	0.00	0	0.00	1980	0
4-Speed Automatic	0.045	225.00	0.00	30	0.00	1980	0.05
5-Speed Automatic	0.065	325.00	0.00	40	0.00	1995	0.07
CVT	0.100	250.00	0.00	20	0.00	1995	0.07
6-Speed Manual	0.020	100.00	0.00	30	0.00	1991	0.05
Electronic Transmission I	0.005	20.00	0.00	5	0.00	1988	0
Electronic Transmission II	0.015	40.00	0.00	5	0.00	1998	0
Roller Cam	0.020	16.00	0.00	0	0.00	1987	0
OHC 4	0.030	100.00	0.00	0	0.00	1980	0.20
OHC 6	0.030	140.00	0.00	0	0.00	1980	0.20
OHC 8	0.030	170.00	0.00	0	0.00	1980	0.20
4C/4V	0.080	240.00	0.00	30	0.00	1988	0.45
6C/4V	0.080	320.00	0.00	45	0.00	1991	0.45
8C/4V	0.080	400.00	0.00	60	0.00	1991	0.45
Cylinder Reduction	0.030	-100.00	0.00	-150	0.00	1988	-0.10
4C/5V	0.100	300.00	0.00	45	0.00	1998	0.55
Turbo	0.050	500.00	0.00	80	0.00	1980	0.45
Engine Friction Reduction I	0.020	20.00	0.00	0	0.00	1987	0
Engine Friction Reduction II	0.035	50.00	0.00	0	0.00	1996	0
Engine Friction Reduction III	0.050	90.00	0.00	0	0.00	2006	0
Engine Friction Reduction IV	0.065	140.00	0.00	0	0.00	2016	0
VVT I	0.080	140.00	0.00	40	0.00	1998	0.10
VVT II	0.100	180.00	0.00	40	0.00	2008	0.15
Lean Burn	0.100	150.00	0.00	0	0.00	2012	0
Two Stroke	0.150	150.00	0.00	-150	0.00	2004	0
TBI	0.020	40.00	0.00	0	0.00	1982	0.05
MPI	0.035	80.00	0.00	0	0.00	1987	0.10
Air Pump	0.010	0.00	0.00	-10	0.00	1982	0
DFS	0.015	15.00	0.00	0	0.00	1987	0.10
Oil %w-30	0.005	2.00	0.00	0	0.00	1987	0
Oil Synthetic	0.015	5.00	0.00	0	0.00	1997	0
Tires I	0.010	16.00	0.00	0	0.00	1992	0
Tires II	0.020	32.00	0.00	0	0.00	2002	0
Tires III	0.030	48.00	0.00	0	0.00	2012	0
Tires IV	0.040	64.00	0.00	0	0.00	2018	0
ACC I	0.005	15.00	0.00	0	0.00	1992	0
ACC II	0.010	30.00	0.00	0	0.00	1997	0
EPS	0.015	40.00	0.00	0	0.00	2002	0
4WD Improvements	0.030	100.00	0.00	0	-0.05	2002	0
Air Bags	-0.010	300.00	0.00	35	0.00	1987	0
Emissions Tier I	-0.010	150.00	0.00	10	0.00	1994	0
Emissions Tier II	-0.010	300.00	0.00	20	0.00	2003	0
ABS	-0.005	300.00	0.00	10	0.00	1987	0
Side Impact	-0.005	100.00	0.00	20	0.00	1996	0
Roof Crush	-0.003	100.00	0.00	5	0.00	2001	0
Increased Size/Wt.	-0.033	0.00	0.00	0	0.05	1991	0
GDI/4-cyl	0.170	1000.00	0.00	0	0.00	2005	0
GDI/6-cyl	0.170	1200.00	0.00	0	0.00	2005	0
Gasoline Hybrid	0.450	0.00	75.00	0	0.05	2005	0

Source: Energy and Environmental Analysis, *NEMS Fuel Economy Model LDV High Technology Update, Final Documentation,* prepared for *Energy Information Administration, (June 1998).*

Notes and Sources

[29] U.S. Department of Transportation, National Highway Traffic and Safety Administration, Mid-Model Year Fuel Economy Reports from Automanufacturers, (1997).

[30] Goldberg, Pinelopi Koujianou, Product Differentiation and Oligopoly In International Markets: The Case of The U.S. Automobile Industry," Econometrica, Vol. 63, No.4 (July, 1995), 891-951.

[31] Maples, John D., The Light-Duty Vehicle MPG Gap: Its Size Today and Potential Impacts in the Future, University of Tennessee Transportation Center, Knoxville, TN, (May 28, 1993, Draft); Decision Analysis Corporation of Virginia, Fuel Efficiency Degradation Factor, Final Report, Prepared for Energy Information Administration (EIA), (Vienna, VA, August 3, 1992); U.S. Department of Transportation, Federal Highway Administration, New Perspectives in Commuting, (Washington, DC, July 1992); U.S. Department of Transportation, Federal Highway Administration, Highway Statistics 1997, FHWA-PL-98-020, (Washington, DC, November 1, 1998); and Green, Tamara, "Re-estimation of Annual Energy Outlook 2000 Degradation Factors," prepared for the Energy Information Administration, unpublished paper, August 18, 1999, Washington, D.C.

[32] U.S. Department of Transportation, op.cit., Note 31.

[33] Decision Analysis Corporation of Virginia, NEMS Transportation Sector Model: Re-estimation of VMT Model, Prepared for Energy Information Administration (EIA), (Vienna, VA, June 30, 1995).

[34] Energy Information Administration, Describing Current and Potential Markets for Alternative Fuel Vehicles, DOE/EIA-0604(96), (Washington, DC, March 1996).

[35] Energy Information Administration, Alternatives to Traditional Transportation Fuels 1996, DOE/(EIA-0585(96), (Washington, DC, December 1997).

[36] Bobbit, The Fleet Fact Book, Redondo Beach, (California, 1995).

[37] U.S. Department of Commerce, Bureau of Census, Truck Inventory and Use Survey 1992, TC-92-T-52, (Washington, DC, May 1995).

[38] U.S. Department of Energy, Office of Policy, Assessment of Costs and Benefits of Flexible and Alternative Fuel Use in the U.S. Transportation Sector, Technical Report Fourteen: Market Potential and Impacts of Alternative-Fuel Use in Light-Duty Vehicles: A 2000/2010 Analysis, (Washington, DC, 1995).

[39] California Resources Board, Proposed Regulations for Low-Emission Vehicles and Fuels, Staff Report, (August 13, 1990).

[40] Oak Ridge National Laboratory, Fleet Vehicles in the United States: Composition, Operating Characteristics, and Fueling Practices, Prepared for the Department of Energy, Office of Transportation Technologies and Office of Policy, Planning, and Analysis, (Oak Ridge, TN, May

Notes and Sources

1992).

[41] *Ibid.*

[42] Energy Information Administration, *op.cit.*, Note 34.

[43] Energy Information Administration, *op.cit.*, Note 35.

[44] U.S. Department of Energy, Office of Energy Efficiency and Renewable Energy, prepared by Interlaboratory Working Group, Scenarios of U.S. Carbon Reductions: Potential Impacts of Energy Technologies by 2010 and Beyond, (Washington DC, 1998).

[45] U.S. Department of Energy, Office of Transportation Technologies and Energy Efficiency and Renewable Energy, Alternative-Fuel Vehicle Model, 1998; and Energy and Environmental Analysis, Changes to the Fuel Economy Module, Final Report, Prepared for Energy Information Administration (EIA), (June 1998).

[46] U.S. Department of Energy, Office of Transportation Technologies, and Argonne National Laboratory, National Alternative Fuel Vehicle Survey, Draft, August 21, 1998.

[47] Energy Information Administration, *op.cit.*, Note 34.

[48] Energy Information Administration, *op.cit.*, Note 35.

[49] Decision Analysis Corporation of Virginia, Re-estimation of Freight Adjustment Coefficients, Report Prepared for Energy Information Administration (EIA), (February 28, 1995).

[50] Reebie Associates, TRANSEARCH Freight Commodity Flow Database, (Greenwich, CT, 1992).

[51] U.S. Department of Commerce, Bureau of Census, *op.cit.*, Note 37.

[52] U.S. Department of Energy, Office of Heavy Vehicle, Technologies (OHVT), OHVT Technology Roadmap, DOE/OSTI-11690, (October 1997).

[53] Energy Information Administration, State Energy Data Report 1996, DOE/EIA-0214(96), (Washington, DC, December 1998).

[54] Decision Analysis Corporation of Virginia, *op.cit.*, Note 49.

[55] Reebie Associates, *op.cit.,* Note 50.

[56] Argonne National Laboratory, Transportation Energy Demand Through 2010, (Argonne, IL, 1992).

[57] U.S. Department of Transportation, Federal Railroad Administration, 1989 Carload Waybill Statistics; Territorial Distribution, Traffic and Revenue by Commodity Classes, (September 1991 and prior issues).

[58] Energy Information Administration, *op.cit.*, Note 53.

Notes and Sources

[59] Decision Analysis Corporation of Virginia, *op.cit.*, Note 49.

[60] Reebie Associates, *op.cit.*, Note 50.

[61] Army Corps of Engineers, Waterborne Commerce of the United States, (Waterborne Statistics Center: New Orleans, LA, 1993).

[62] Energy Information Administration, *op.cit.*, Note 53.

[63] Transportation Research Board, Forecasting Civil Aviation Activity: Methods and Approaches, Appendix A, Transportation Research Circular Number 372, (June 1991).

[64] Decision Analysis Corporation of Virginia, Re-estimation of NEMS Air Transportation Model, Prepared for the Energy Information Administration (EIA), (Vienna, VA, 1995).

[65] Air Transport Association of America, Air Travel Survey, (Washington DC, 1990).

[66] U.S. Department of Transportation, Air Carrier Traffic Statistics Monthly, (December 1996).

[67] U.S. Department of Transportation, Federal Aviation Administration, FAA Aviation Forecasts Fiscal Years 1996-2008, (Washington, DC, March 1997, and previous editions).

[68] *Ibid*.

[69] Oak Ridge National Laboratory, Energy Efficiency Improvement of Potential Commercial Aircraft to 2010, ORNL-6622, (Oak Ridge, TN, June 1990), Oak Ridge National Laboratory, Air Transport Energy Use Model, Draft Report, (Oak Ridge, TN, April 1991).

[70] Ibid.

[71] Energy Information Administration, *op.cit.*, Note 34.

[72] Energy Information Administration, *op.cit.*, Note 35.

[73] Bobbit, *op.cit.*, Note 36.

[74] U.S. Department of Commerce, Bureau of Census, *op.cit.*, Note 37.

[75] U.S. Department of Energy, Office of Policy, *op.cit.*, Note 38.

[76] U.S. Department of Energy, Office of Policy, *op.cit.*, Note 38.

[77] California Air Resources Board, Proposed Regulations for Low Emission Vehicles and Clean Fuels, Staff Report, (August 13, 1990).

Notes and Sources

[78] State of California Air Resources Board, Staff Report: Initial Statement of Reasons, Proposed Amendments to California Exhaust and Evaporative Emissions Standards and Test Procedures For Passenger Cars, Light-Duty Trucks and Medium-Duty Vehicles -"LEV II" and Proposed Amendments to California Motor Vehicle Certification, Assembly-Line and In-Use Test Requirements -"CAP 2000," Mobile Source Control Division, El Monte, CA, September 18, 1998.

[79] Http://www.epa.gov/fedrgstr/EPA-AIR/1999/May/Day-13/a11384a.htm.

[80] U.S. EPA, Office of Mobile Sources, Exhaust Emission Certification Standards, EPA 420-B-98-001, March 24, 1998.

[81] Http://www.epa.gov/oms/tr2home.htm.

[82] National Research Council, Review of the Research Program of the Partnership for a New Generation of Vehicles: Fifth Report, National Academy Press, Washington, D.C., 1999.

[83] U.S. Department of Energy, Office of Energy Efficiency and Renewables, Office of Transportation Technologies, OTT Program Analysis Methodology: Quality Metrics 2000, (November 1, 1998).

[84] DeCicco, John, and Marc Ross, An Updated Assessment of the Near-Term Potential for Improving Automotive Fuel Economy, American Council for an Energy Efficient Economy, November 1993.

[85] U.S. Department of Energy, Office of Heavy Vehicle, Technologies, op.cit., Note 52.

[86] National Research Council, Aeronautics and Space Engineering Board, 1992. Aeronautical Technologies for the Twenty-First Century, National academy Press, Washington, D.C.

Electricity Market Module

The NEMS Electricity Market Module (EMM) represents the planning, operations, and pricing of electricity in the United States. It is composed of four primary submodules—electricity capacity planning, electricity fuel dispatching, load and demand-side management, and electricity finance and pricing. In addition, nonutility generation and supply and electricity transmission and trade are represented in the planning and dispatching submodules.

Based on fuel prices and electricity demands provided by the other modules of the NEMS, the EMM determines the most economical way to supply electricity, within environmental and operational constraints. There are assumptions about the operations of the electricity sector and the costs of various options in each of the EMM submodules. The major assumptions are summarized below.

Key Assumptions

Capacity Types

Twenty-six capacity types are presented in the EMM (Table 36).

Table 36. Capacity Types Represented in the Electricity Market Module

Capacity Type
Coal Steam pre-1965; Unscrubbed coal - Sulfur dioxide   <=1.20 pounds per million Btu
Coal Steam pre-1965; Unscrubbed coal - Sulfur dioxide   <=3.34 pounds per million Btu
Coal Steam pre-1965; Unscrubbed coal - Sulfur dioxide   >=3.34 pounds per million Btu
Coal Steam post-1965; Unscrubbed coal - Sulfur dioxide   <= 1.20 pounds per million Btu
Coal Steam post-1965; Unscrubbed coal - Sulfur dioxide   <= 3.34 pounds per million Btu
Coal Steam post-1965; Unscrubbed coal - Sulfur dioxide   >=3.34 pounds per million Btu
Coal Steam with Scrubber
New High Sulfur Pulverized Coal with Wet Flue Gas Desulfurization
New Advanced Coal - Integrated Coal Gasification Combined Cycle
Oil/Gas Steam - Oil/Gas Steam Turbine
Combined Cycle - Conventional Gas/Oil Combined Cycle Combustion Turbine
New Advanced Combined Cycle - Advanced Gas/Oil Combined Cycle Combustion Turbine
Combustion Turbine - Conventional Combustion Turbine
Advanced Combustion Turbine - Steam Injected Gas Turbine
Molten Carbonate Fuel Cell
Advanced Nuclear Advanced Light Under Reactor
Conventional Hydropower - Hydraulic Turbine
Pipeline Hydropower - Hydraulic Turbine
Pumped Storage - Hydraulic Turbine Reversible
Geothermal - Dual Flash
Geothermal - Binary
Municipal Solid Waste - Mass Burn
Biomass - Integrated Gasification Combined-Cycle
Solar Thermal - Central Receiver
Solar Photovoltaic - Fixed-Flat Plate
Wind

Source: Energy Information Administration, Office of Integrated Analysis and Forecasting.

New Generating Plant Characteristics

The operational characteristics of new generating technologies are the most important inputs to the electricity capacity planning submodule. The key characteristics for these technologies are summarized in Table 37. These characteristics are used, in combination with fuel price foresight from the NEMS Integrating Module, to compare resource options when new capacity is needed. Heat rates for fossil-fueled technologies decline linearly between 1995 and 2010. The assumptions for nuclear technologies are described later in this section.

The overnight costs listed for each technology in Table 37 are the base costs estimated to build a plant in *Middletown, U.S.A.* Differences in plant costs due to regional distinctions are calculated by applying regional multipliers (Tables 38 and 39) to the cost of labor, factory equipment, and site material for each new generating technology.

Table 37. Cost and Performance Characteristics of New Central Station Electricity Generating Technologies

Technology	Year Available	Size (mW)	Leadtime (Year)	Overnight Capital Costs[1] In 1999 ($1998/kW)	Variable O&M (1998 Mills/kWhr)	Fixed O&M ($1998/kW)	Heatrate First-of-a-kind (Btu/kWhr)	Heatrate Nth-of-a-kind (Btu/kWhr)
Scrubbed Coal New	1997	400	4	1,102	3.33	23.03	9,585	9,087
Integrated Gas Comb Cycle	1997	428	4	1,315	0.79	32.13	8,470	6,968
Gas/Oil Steam Turbine	1997	300	2	1,012	0.51	30.70	9,500	9,500
Conv Gas/Oil Comb Cycle	1997	250	3	449	0.51	15.35	8,030	7,000
Adv Gas/Oil Comb Cycle	1997	400	3	580	0.51	14.23	6,985	6,350
Conv Combustion Turbine	1998	160	2	332	0.10	6.35	11,900	10,600
Adv Combustion Turbine	1997	120	2	465	0.10	9.01	9,700	8,000
Fuel Cells	2001	10	2	2,163	2.05	14.74	6,000	5,361
Advanced Nuclear	2001	600	4	2,390	0.41	56.29	10,400	10,400
Biomass	2001	100	4	1,877	5.32	44.00	9,224	8,219
MSW[2]	1996	30	1	4,424	5.53[2]	0.00	16,000	16,000
Geothermal[3]	1997	50	4	1,621	0.00	85.90	32,391	N/A
Wind	1997	50	3	1,086	0.00	25.92	N/A	N/A
Solar Thermal[4,5]	1997	100	3	3,059	0.00	46.58	N/A	N/A
Photovoltaic[5]	1998	5	2	4,836	0.00	9.82	N/A	N/A

[1]Overnight capital cost (i.e., excluding interest charges) plus project contingencies, excluding regional multipliers (See Tables 38 and 39). These estimates are costs of new projects that are initiated in 1999.

[2]Because municipal solid waste (MSW) does not compete with other technologies in the model, these values are used only in calculating the average costs of electricity.

[3]because geothermal cost and performance parameters are specific for each of the 51 sites in the database, the Nth-of-a-kind capital cost and heatrate are averages for the capacity built in 2000.

[4]Solar thermal is assumed to operate economically only in Electricity Market Module regions 2, 5, and 10-13, that is, West of the Mississippi River, because of its requirement for significant direct, normal insolation.

[5]Capital costs for solar technologies are net of (reduced by) the 10 percent investment tax credit.

O&M = Operation and maintenance.

Sources: Most values are derived by the Energy Information Administration, Office of Integrated Analysis and Forecasting from analysis of reports and discussions with various sources from industry, government, and the National Laboratories, with the

Table 38. Regional Multipliers for New Construction, Fossil-Fueled and Nuclear Generating Technologies

EMM Region	NE, NY	MAAC	STV	MAPP, ECAR MAIN	SPP
Factory Equipment	1.09	1.01	0.95	1.01	1.03
Site Labor	1.33	0.97	0.69	1.03	0.98
Site Material	1.08	0.97	0.93	1.00	1.00
EMM Region	**RA**	**NWP**	**FL**	**CNV**	**ERCOT**
Factory Equipment	1.05	0.99	0.90	1.01	1.02
Site Labor	1.02	1.20	0.70	1.45	0.89
Site Material	1.03	1.00	0.80	1.01	0.98

Note: See Part II, Detailed Tables, Tables 54 through 66 for regional descriptions.

Source: Argonne National Laboratory, *Cost and Performance Database for Electric Power Generating Technologies.*

Table 39. Regional Multipliers for New Construction, Renewable Energy Technologies

EMM	Number/Region	Multiplier
1	ECAR	1.01;
2	ERCOT	1.00; 0.98 for MSW
3	MAAC	1.00; 0.99 for MSW
4	MAIN	1.01
5	MAPP	1.01
6	NY	1.12; 1.16 for MSW
7	NE	1.12; 1.16 for MSW
8	FL	0.86; 0.83 for MSW
9	STV	0.91; 0.87 for MSW
10	SPP	1.02; 1.01 for MSW
11	NWP	1.02; 1.00 for geothermal; 1.05 for MSW
12	RA	1.04; 1.00 for geothermal
13	CNV	1.07; 1.00 for geothermal; 1.13 for MSW

Source: Argonne National Laboratory, *Cost and Performance Database for Electric Power Generating Technologies.*

Representation of Electricity Demand

The annual electricity demand projections from the NEMS demand modules are converted into load duration curves for each of the EMM regions (based on North American Electric Reliability Council regions and subregions) using historical hourly load data. However, unlike traditional load duration curves where the demands for an entire period would be ordered from highest to lowest, losing their chronological order, the load duration curves in the EMM are segmented into nine different time slices (Table 40). The time periods shown were mainly chosen to accommodate intermittent generating technologies (i.e., solar and wind facilities) and demand-side management programs.

Table 40. Load Segments for the Electricity Market Module

Season	Months	Period	Hours
Summer	June-September	Daytime	0700-1800
		Morning/Evening	0500-0700, 1800-2400
		Night	0000-0500
Winter	December-March	Daytime	0800-1600
		Morning/Evening	0500-0800, 1600-2400
		Night	0000-0500
Off-peak	April-May	Daytime	0700-1700
	October-November	Morning/Evening	0500-0700, 1700-2400
		Night	0000-0500

Note: Both the summer and winter peak periods are represented by 2 vertical slices each (a peak slice and an off-peak slice). The remaining 7 periods are represented by 1 vertical slice each, resulting in a total of 11 vertical slices.

Source: Energy Information Administration, Office of Integrated Analysis and Forecasting.

Reserve margins—the percentage of capacity required in excess of peak demand needed for unforeseeable outages—are also assumed for each regulated EMM region. A thirteen percent reserve margin is assumed for MAPP, STV and SPP, nine percent for FL, and fifteen percent for NWP. In the other regions, which are assumed to be fully or partially deregulated, the EMM determines the reserve margin by equating the marginal cost of capacity and the cost of unserved energy.

Fossil Fuel-Fired Steam Plant Maintenance/Retirement

Fossil-fired steam plant retirements are calculated endogenously within the model. Fossil plants are retired when it is no longer economical to continue running them. Each year, the model determines whether the market price of electricity is sufficient to support the continued operating of existing plants. If the expected revenues from these plants are not sufficient to cover the annual going forward costs - mainly fuel and operations and maintenance costs - the plant will be retired if the overall cost of producing electricity can be lowered by building new replacement capacity.

Nuclear Power Plant Orders and Retirements

There are no nuclear units currently under construction in the United States. New nuclear capacity will be added if the costs are competitive with other generating technologies.

It is assumed that older nuclear power plants will incur aging related expenditures in the form of either increased capital costs, decreases in performance and/or increased maintenance expenditures to maintain a given level of performance. The decision to either incur the aging-related costs of the unit or retire the unit is based on the relative economics of the alternatives. In AEO2000, the retirement decision for each nuclear unit is evaluated every 10 years, starting after roughly 30 years of operation. An assumption is made regarding the capital costs required to operate an additional 10 years beyond the point of evaluation. In the reference case, the aging-related capital costs over ten years are assumed to be $150 milliion at 30 years, an additional $175 million at 40 years and $250 million at 50 years, where dollar amounts are based on an average plant size of 1,000 megawatts. The investment cost is assumed to be recovered over ten years, and an annual payment is calculated. If the combined operating costs and annual capital payment costs are lower than the cost of building new capacity, then the nuclear unit continues to operate another 10 years, until the next evaluation.

Plants that have recently incurred a major retrofit are assumed not to incur aging-related expenses between 30 and 40 years and only one-third of the costs from 40 to 50 years. Additionally, the aging-related cost assumptions are adjusted downward for the newest vintage of nuclear reactors, to reflect improvements in construction and design.

Interregional Electricity Trade

Both firm and economy electricity transactions among utilities in different regions are represented within the EMM. In general, firm power transactions involve the trading of capacity and energy to help another region satisfy its reserve margin requirement, while economy transactions involve energy transactions motivated by the marginal generation costs of different regions. The flow of power from region to region is constrained by the existing and planned capacity limits as reported in the NERC and WSCC Summer and Winter Assessment of Reliability of Bulk Electricity Supply in North America. Known firm power contracts are obtained from NERC's *Electricity Supply and Demand Database 1998*. They are locked in for the term of the contract and then phased out through 2020. In addition, in certain regions where data show an established commitment to build plants to serve another region, new plants are permitted to be built to serve the other region's needs. This option is available to compete with other resource options.

Economy transactions are determined in the dispatching submodule by comparing the marginal generating costs of adjacent regions in each time slice. If one region has less expensive generating resources available in a given time period (adjusting for transmission losses and transmission capacity limits) than another region, the regions are allowed to exchange power.

International Electricity Trade

Two components of international firm power trade are represented in the EMM—existing and planned transactions, and unplanned transactions. Existing and planned transactions are obtained from the North American Electric Reliability Council's *Electricity Supply and Demand Database 1998*. Unplanned firm power trade is represented by competing Canadian supply with U.S. domestic supply options. Canadian supply is represented via supply curves using cost data from the Department of Energy report *Northern Lights: The Economic and Practical Potential of Imported Power from Canada*, (DOE/PE-0079).

International economy trade is determined endogenously based on surplus energy expected to be available from Canada by region in each time slice. Canadian surplus energy is determined using Canadian electricity supply and demand projections as reported in the Canadian National Energy Board report *Energy Supply and Demand 2025*.

Electricity Finance and Pricing

The reference case assumes a transition to full competitive pricing in California, New York, the New England and Mid-Atlantic Area Council, and Texas. In addition electricity prices in the East Central Area Reliability Council, the Mid-American Interconnected Network (Illinois, plus parts of Missouri Michigan and Wisconsin), and the Rocky Mountain Power Area/ Arizona are assumed to be partially competitive. Some of the States in each of these regions have not taken action to deregulate their pricing of electricity, and in those States prices are assumed to continue to be based on traditional cost-of-service pricing. The reference case assumes that: in California, the price of electricity will remain constant between 1996 and 2001 for commercial and industrial consumers while residential customers will enjoy a 10 percent reduction in current prices starting in 1998; the market will transition from a regulated to a competitive market between 2002 and 2007; and California markets will be fully competitive by 2008. Similarly, in the other competitive regions, the transition period is assumed to occur over a ten-year period beginning in 1999 or 2000.

The price of electricity to the consumer is comprised of the price of generation, transmission and distribution. Transmission and distribution are considered to remain regulated in the *AEO*; that is, the price of transmission and distribution is based on the average cost for each customer class. In the competitive regions, the generation component of price is based on marginal cost, which is defined as the cost of the last (or most expensive) unit dispatched. The marginal cost includes fuel, operating and maintenance, taxes, and a reliability price adjustment, which represents the value of capacity in periods of high demand. Therefore, the price of electricity in the regulated regions consists of the average cost of generation, transmission, and distribution for each customer class. The price of electricity in the five regions with a competitive generation market consists of the marginal cost of generation summed with the average costs of transmission and distribution.

In recent years, the move towards competition in the electricity business has led utilities to make efforts to reduce costs to improve their market position. These cost reduction efforts are reflected in utility operating data reported to the Federal Energy Regulatory Commission (FERC) and trends evidenced there have been incorporated in the *AEO2000*. The key trends are discussed below:

- Reduced General and Administrative Expenses (G&A) - Over the 1990 through 1994 period, utilities have reduced their employment by 65,000, a reduction of nearly 3 percent annually. This trend has been incorporated by reducing G&A expenditures at a rate of 2.5 percent annually through 2005. No further reductions are assumed to occur after 2005.

- Reduced Fossil Plant Operations Expenditures (O&M) - Again, over the 1990 through 1994 period, utility fossil plant operation and maintenance costs (all operating costs other than fuel) fell at a rate of nearly 3 percent annually. As with G&A, this trend has been incorporated by reducing fossil O&M expenditures at a rate of 2.5 percent annually through 2005. No further reductions are assumed to occur after 2005.

- Reduced Nuclear Operations and Maintenance Expenditures - In the *AEO2000* nuclear O&M expenditures are reduced over time to reflect the impact of older more expensive plants retiring in the later years of the forecast. In 2020 nuclear capacity is 40 gigawatts below the 1998 level and nuclear O&M expenditures are reduced 5 percent to reflect this.

Demand-Side Management

Improvements in energy efficiency induced by rising energy prices, new appliance standards, and utility demand-side management programs are represented in the end-use demand models. Appliance choice decisions are a function of the relative costs and performance characteristics of a menu of technology options. In 1997, utilities reported spending over $1.64 billion on demand-side management programs.[87]

Fuel Price Expectations

Capacity planning decisions in the EMM are based on a lifecycle cost analysis over a 20-year period. This requires foresight assumptions for fuel prices. Expected prices for coal, natural gas, and oil are derived using adaptive expectations, in which future prices are extrapolated from recent historical trends.[88] For each projection year, coal prices are assumed to decrease one percent annually from that year's projected price until the end of the subsequent 20 year period. For each oil product, future prices are estimated by applying a constant markup to an external forecast of world oil prices. The markups are calculated by taking the differences between the regional product prices and the world oil price for the previous forecast year. For natural gas, expected wellhead prices are based on a nonlinear function that relates the expected price to the expected cumulative domestic gas production. Delivered prices are developed by applying a constant markup, which represents the difference between the delivered and wellhead prices from the prior forecast year.

The approach for natural gas was developed to have the following properties:

1. The natural gas wellhead price should be upward sloping as a function of cumulative gas production.

2. The rate of change in wellhead prices should increase as fewer economical reserves remain to be discovered and produced.

The approach assumes that at some point in the future a given target price, PF, results when cumulative gas production reaches a given level, QF. The target values for PF and QF were assumed to be $6.00 per thousand cubic feet (1995 dollars) and 2000 trillion cubic feet, respectively. Gas hydrates are included in the resource base. The future annual production is assumed to be constant at the prior year's level.

The expected wellhead gas price equation is of the following form:

$$P_k = A * Q_k^{0.75} + B$$

where P is the wellhead price for year k, Q_k is the cumulative production from 1991 to year k, and A and B are determined each year such that the price equation will intersect the future target point (PF, QF).

Technological Optimism and Learning

Overnight costs for each technology are calculated as a function of regional construction parameters, project contingency, and technological optimism and learning factors. For each generating technology available for new capacity in a region, the overnight cost used by the model is calculated using the base cost for the technology from Table 37 and the technological and learning parameters from Table 41.

The learning function has the nonlinear form:

$$OC(C) = a*C^{-b},$$

Where C is the cumulative capacity for the technology.

The progress ration (*pr*) is defined by speed of learning (e.g., how much costs decline for every doubling of capacity). The reduction in capital cost for every doubling of cumulative capacity (f) is an exogenous parameter input for each technology Table 41. Consequently, the progress ratio and *f* are related by:

$$pr = 2^{-b} = (1 - f)$$

The parameter "b" is calculated by $(b = -(\ln(1-f)/\ln(2))$. The parameter "a" can now be found from initial conditions. That is,

$$a = OC(C0)/C0^{-b}$$

Where C0 is the cumulative initial capacity. Thus, once the rates of learning (f) and the cumulative capacity (C0) are known for each interval, the corresponding parameters (a and b) of the nonlinear function are known. The overnight costs (OC) computed in this manner are then adjusted to account for technological optimism.

Table 41. Technological Optimism and Learning Parameters for New Generating Technologies

Technology	Initial Optimism Factor	Period[1] Learning Factor	Period[2] Learning Factor	Period[3] Learning Factor	Vintage	Period[1] Doublings	Period[2] Doublings
Scrubbed Pulverized Coal	1.00	0.1	0.05	0.01	Conventional	3	5
Integrated Gas Comb Cycle	1.00	0.1	0.05	0.01	Evolutionary	3	5
Gas/Oil Steam Turbine	1.00	0.1	0.05	0.01	Conventional	3	5
Conv Gas/Oil Comb Cycle	1.00	0.1	0.05	0.01	Conventional	3	5
Adv Gas/Oil Comb Cycle	1.00	0.1	0.05	0.01	Evolutionary	3	5
Conv Combustion Turbine	1.00	0.1	0.05	0.01	Conventional	3	5
Adv Combustion Turbine	1.00	0.1	0.05	0.01	Evolutionary	3	5
Fuel Cells	1.16	0.1	0.05	0.01	Revoluntionary	3	5
Advanced Nuclear	1.19	0.1	0.05	0.01	Revoluntionary	3	5
Biomass (Wood)	1.19	0.1	0.05	0.01	Revoluntionary	3	5
Municipal Solid Waste[1]	1.00	0.1	0.05	0.01	Conventional	3	5
Geothermal	1.00	0.1	0.05	0.01	Evolutionary	3	5
Wind	1.00	0.1	0.05	0.01	Evolutionary	3	5
Solar Thermal	1.19	0.1	0.05	0.01	Revoluntionary	3	5
Photovoltaic	1.12	0.2	0.05	0.01	Revolutionary	3	5

[1] Municipal Solid Waste does not compete in the model with other technologies; thereby, Optimism and Learning Factors are not computed.

Source: Energy Information Administration, Office of Integrated Analysis and Forecasting.

In *AEO2000*, capital costs for all new electricity generating technologies (fossil, nuclear, and renewable) decrease in response to foreign and domestic experience. Foreign units of new technologies are assumed to contribute to reductions in capital costs for units that are installed in the United States to the extent that (1) the technology characteristics are similar to those used in U.S. markets, (2) the design and construction firms and key personnel compete in the U.S. market, (3) the owning and operating firm competes actively in the U.S., market, and (4) there exists relatively complete information about the status of the associated facility. If the new foreign units do not satisfy one or more of these requirements, they are given a reduced weight or not included in the domestic learning effects calculation.

International Learning. For *AEO2000*, capital costs for all new fossil-fueled electricity generating technologies decrease in response to foreign as well as domestic experience, to the extent that the new plants reflect technologies and firms also competing in the United States. *AEO2000* includes 3,887 megawatts of advanced coal gasification combined cycle capacity and 15,177 megawatts of advanced combined cycle natural gas capacity to be built outside the United States after 1999 and through 2003.

Legislation

Clean Air Act Amendments of 1990 (CAAA90)

It is assumed that electricity producers comply with the CAAA90, which mandate a limit of 9.48 million short tons of sulfur dioxide emissions by 2000 and 8.95 million tons by 2010. Utilities are assumed to comply with the limits on sulfur emissions by retrofitting units with flue gas desulfurization (FGD) equipment, transferring or purchasing sulfur emission allowances, operating high-sulfur coal units at a lower capacity utilization rate, or switching to low-sulfur fuels. The costs for FGD equipment average approximately $195 per kilowatt, in 1988 dollars, although the costs vary widely across the regions. It is also assumed that the market for trading emission allowances is allowed to operate without regulation and that the States do not further regulate the selection of coal to be used.

Utilities are assumed to comply with the mandates set forth in the CAAA90 with respect to the SO_2 and NO_x standards. It is assumed that utilities will comply with CAAA90 and reduce their emissions of sulfur dioxide (SO_2) by 10 million tons over the forecast period. Consequently, the forecast assumes that the cost associated with purchasing an SO_2 allowance (dollars per ton of SO_2) is equivalent to the marginal cost of compliance (dollars per ton of SO_2 removed).

As specified in the CAAA90, EPA has developed a two-phase NOx program, with the first set of standards for existing coal plants taking force in 1996 while the second set is to be implemented in 2000 (Table 42). Dry bottom wall-fired, and tangential fired boilers, the most common boiler types, referred to as Group 1 Boilers, were required to make significant reductions beginning in 1996 and further reductions in 2000. Relative to their uncontrolled emission rates, which range roughly between 0.6 and 1.0 pounds per million Btu, they are required to make reductions of between 25 and 50 percent to meet the Phase I limits and further reductions to meet their Phase II limits. The EPA did not impose limits on existing oil and gas plants, but some states

Table 42. NO$_x$ Emissions Standards
(Pounds per million Btu)

Boiler Type	# Boilers	Phase I Limit	Phase II Limit
Group 1 Boilers			
Dry Bottom Wall-Fired	284	0.50	0.45
Tangential	296	0.45	0.38
Group 2 Boilers			
Cell Burners	35	NA	0.68
Cyclones	88	NA	0.94
Wet Bottom Wall-Fired	38	NA	0.86
Vertically Fired	29	NA	0.80
Fluidized Bed	5	NA	0.29

NA = Not Applicable

Source: Environmental Protection Agency, Nitrogen Oxide Emission Reduction Program

have No$_x$ regulations. All new fossil units are required to meet standards. In pounds per million Btu, these limits are 0.11 for conventional coal, 0.02 for advanced coal, 0.02 for combined cycle, and 0.08 for combusion turbines. All of these No$_x$ limits are incorporated in the NEMS.

Energy Policy Act of 1992 (EPACT)

The provisions of the EPACT include revised licensing procedures for nuclear plants and the creation of exempt wholesale generators (EWGs).

EPACT allows the issuance of a combined construction and operating license for nuclear plants; however, it also allows for a post-construction hearing and judicial review. The uncertainty associated with waste, regulatory, and financial issues is sufficiently large to require their resolution or some manner of financial protection for investors before investments in nuclear power would take place. Unresolved, these conditions would lead to investments in alternative capacity additions or a delay in capital investment. Therefore, no newly ordered nuclear plants are assumed to become operational by 2020.

EPACT reformed the Public Utility Holding Company Act of 1935 (PUHCA). Prior to the passage of EPACT, PUHCA required that utility holding companies register with the Securities and Exchange Commission (SEC) and restricted their business activities and corporate structures.[89]

Entities that wished to develop facilities in several States were regulated under PUHCA. To avoid the stringent SEC regulation, nonutilities had to limit their development to a single State or limit their ownership share of projects to less than 10 percent. EPACT changed this by creating a class of generators that, under certain conditions, are exempt from PUHCA restrictions. These EWGs can be affiliated with an existing utility (affiliated power producers) or independently owned (independent power producers). In general, subject to State commission approval, these facilities are free to sell their generation to any electric utility, but they cannot sell to a retail consumer. These EWGs are represented in NEMS.

Climate Change Action Plan

As a result of the Climate Challenge Program (CCAP) many utilities have announced efforts to voluntarily reduce their greenhouse gas emissions between now and 2000. These efforts cover a wide variety of programs including increasing DSM investments, repowering (fuel-switching) of fossil plants, restarting of nuclear plants that have been out-of-service, planting trees, and purchasing emission offsets from international sources. To the degree possible, each one of the participation agreements was examined to determine if the commitments made were addressed in the normal reference case assumptions or whether they were addressable in NEMS. Programs like tree planting and emission offset purchasing are not addressable in NEMS. With regard to the other programs, they are, for the most part, captured in NEMS. For example, utilities annually report to EIA their plans (over the next 5 years) to bring a plant back on line, repower a plant, life extend a plant, cancel a previously planned plant, build a new plant, or switch fuel at a plant. Additionally, reduced transmission losses due to improved transformer efficiencies are incorporated. These data are inputs to NEMS. Thus, programs that would affect these areas are reflected in NEMS input data. However, because many of the agreements do not identify the specific plants where action is planned, it is not possible to determine which of the specified actions, together with their greenhouse gas emission savings, should be attributed to the Climate Challenge Program and which are just the result of normal business operations.

FERC Orders 888 and 889

FERC has issued two related rules (Orders 888 and 889) designed to bring low cost power to consumers through competition, ensure continued reliability in the industry, and provide for open an equitable transmission services by owners of these facilities. Specifically, Order 888 requires open access to the transmission grid currently owned and operated by utilities. The transmission owners must file nondiscriminatory tariffs that offer other suppliers the same services that the owners provide for themselves. Order 888 also allows these utilities to recover stranded costs (investments in generating assets that are unrecoverable due to consumers selecting another supplier). Order 889 requires utilities to implement standards of conduct and a Open Access Same-time Information System (OASIS) through which utilities and non-utilities can receive information regarding the transmission system. Consequently, utilities are expected to functionally or physically unbundle their marketing functions from their transmission functions.

These orders are represented in the EMM by assuming that the debt/equity financing structure for new technologies is the same for utilities and nonutilities.

Electricity and Renewable Technology Cases

High Electricity Demand Case

The *high electricity demand case* assumes that electricity demand grows at 2.0 percent annually between 1998 and 2020. In the reference case, electricity demand is projected to grow 1.4 percent annually between 1998 and 2020. No attempt was made to determine the changes needed in the end-use sectors to result in the stronger demand growth.

The *high electricity demand case* is a partially integrated run. The end-use demand modules are not operated, but all of the electricity end-use demands from the reference case are multiplied by the same factor to achieve the higher growth rate. Using the higher electricity demand and all other reference case demand projections as inputs, the EMM, Macroeconomic Activity, Petroleum Marketing, International

Energy, Oil and Gas, Natural Gas Transmission and Distribution, Coal Market, and Renewable Fuels Modules are allowed to interact.

Low and High Fossil Cases

The *low fossil case* assumes that the costs of advanced generating technologies (integrated coal-gasification combined-cycle, advanced natural gas combined-cycle and turbines, and fuel cells) will remain at the base cost during the projection period. Capital costs of conventional generating technologies are the same as those assumed in the reference case (Table 43). In the *high fossil case*, efficiencies of advanced fossil generating technologies are higher than the reference case, based on discussions with the Department of Energy, Office of Fossil Energy, while efficiencies of conventional technologies are the same as used in the reference case.

The *low and high fossil* runs are partially-integrated runs, i.e., the Macroeconomic Activity, Petroleum Market, International Energy, and end-use demand modules use the reference case values and are not effected by changes in generating capacity mix. Conversely, the Oil and Gas Supply, Natural Gas Transmission and Distribution, Coal Market, and Renewable Fuels Modules are allowed to interact with the EMM in the *low and high fossil cases*.

Low and High Nuclear Cases

Two side cases were developed with different assumptions regarding the capital investments, which change the retirement decisions. In the low nuclear case the capital investment is increased by $50 million at each decisions point. In addition, the adjustments for the new plants were removed, making these units face higher capital investments. The high nuclear case assumes that no additional investment is needed during the first 40 years of operation, and that capital expenditures are reduced by $100 to $125 million after 40 years.

The *low and high nuclear cases* are partially-integrated model runs, i.e., the Macroeconomic Activity, Petroleum Market, and International Energy modules use the reference case outputs and are not affected by changes in nuclear capacity. Conversely, the Oil and Gas Supply, Natural Gas Transmission and Distribution, Coal Market, and Renewable Fuels Modules interact with the EMM in the high and low nuclear cases.

High Renewables Case

For the *high renewables case*, EIA incorporates approximations of renewable energy technology characterizations prepared jointly by the U.S. Department of Energy and the Electric Power Research Institute, technology assumptions of lower capital and operating costs, and higher efficiencies (capacity factors) for new renewable energy generating technologies than used in the reference case.[90] EIA also assumed that the yields for energy crops grown on pasture and crop land are nearly 20 percent higher than in the reference case. All other technologies and other NEMS modeling characteristics remain unchanged from the reference case (Table 44).

Renewable Portfolio Standard Case

Three different cases involving minimum requirements for nonhydroelectric renewable generation are represented. Each case uses the requirement of 7.5 percent of electricity sales by 2010, as specified in the Administration's proposed Comprehensive Electricity Competition Act. The *RPS with cap and sunset case* includes a 1.5 cents per kilowatthour cap on the renewable credit price. That is, suppliers can purchase credits rather than generate electricity from renewables if the market price for credits exceeds this cap. It also contains a sunsetting provision, in which the required share is held constant through 2015 and then expires thereafter. The *RPS with cap, no sunset case* incorporates the price cap, but removes the sunsetting provision. *RPS no cap, no sunset case* contains neither the cap or sunset options and therefore requires that the standard be achieved through 2020.

Competitive Pricing Cases

There are three competitive pricing cases that differ in their assumptions on natural gas supply technology development. The *mid-price gas case* uses the reference case assumptions on natural supply technology. The *competitive pricing case with low gas prices* incorporates the oil and natural gas supply technology assumptions from the oil and gas rapid technology case. Similarly, the *competitive pricing case with high gas prices* incorporates the oil and natural gas supply technology assumptions from the oil and gas slow technology case.

Each competitive pricing case assumes that all regions of the country will gradually move toward marginal-cost-based pricing for generation services. Competitive pricing for generation services is phased in over 10 years by computing a weighted average of the traditional average-cost-based price and a price based on marginal costs. The weighting factor changes over time—initially weighting the average-cost-based price more heavily, then decreasing the weight over the phase-in period—until the price is based solely on marginal costs. Prices in two regions, NWP and STV, are weighted to reflect the assumption that public power would still be priced at average costs.

It is also assumed that some consumers will be able to respond to time-of-use pricing by altering their demand patterns. Through "load shifting", consumer can reduce usage during a peak period, when prices are high and supply is tight, and shift that usage to an off-peak period.

Table 43. Cost and Performance Characteristics for Fossil-Fueled Generating Technologies: Three Cases

Technology	Base Costs Reference (1998$/kW)	Overnight Costs			Heat Rate		
		Reference (1998$/kW)	High Fossil (1998$/kW)	Low Fossil (1998$/kW)	Reference (Btu/kWh)	High Fossil (Btu/kWh)	Low Fossil (Btu/kWh)
Pulverized Coal	$1,102						
2000		$1,102	$1,027	$1,104	9,419	8,979	9,419
2005		$1,088	$ 909	$1,101	9,253	8,124	9,253
2010		$1,075	$ 808	$1,081	9,087	7,582	9,087
2015		$1,062	$ 808	$1,065	9,087	7,582	9,087
2020		$1,049	$ 808	$1,049	9,087	7,582	9,087
Integrated Coal Gasification Combined-Cycle	$1,315						
2000		$1,239	$1,111	$1,623	7,969	7,937	8,470
2005		$1,220	$1,010	$1,623	7,469	7,403	8,470
2010		$1,191	$1,010	$1,623	6,968	6,438	8,470
2015		$1,140	$1,010	$1,623	6,968	5,687	8,470
2020		$1,091	$1,010	$1,623	6,968	5,687	8,470
Conv. Comb.-Cycle	$ 332						
2000		$330	$ 317	$ 314	11,467	11,467	11,467
2005		$362	$ 317	$ 307	11,033	11,033	11,033
2010		$324	$ 317	$ 306	10,600	10,600	10,600
2015		$320	$ 317	$ 301	10,600	10,600	10,600
2020		$316	$ 317	$ 298	10,600	10,600	10,600
Adv.Comb.-Cycle	$465						
2000		$464	$ 466	$ 466	9,133	8,865	9,700
2005		$408	$ 378	$ 466	8,567	8,699	9,700
2010		$368	$ 289	$ 466	8,000	8,533	9,700
2015		$357	$ 289	$ 466	8,000	8,533	9,700
2020		$355	$ 289	$ 466	8,000	8,533	9,700
Conv. Comb. Turbine	$449						
2000		$448	$ 427	$ 447	7,687	7,687	7,687
2005		$443	$ 427	$ 427	7,343	7,343	7,343
2010		$438	$ 427	$ 419	7,000	7,000	7,000
2015		$432	$ 427	$ 412	7,000	7,000	7,000
2020		$427	$ 427	$ 408	7,000	7,000	7,000
Adv. Comb. Turbine	$580						
2000		$561	$ 580	$ 580	6,927	6,919	6,985
2005		$527	$ 492	$ 580	6,639	6,204	6,985
2010		$502	$ 404	$ 580	6,350	6,204	6,985
2015		$488	$ 404	$ 580	6,350	4,874	6,985
2020		$477	$ 404	$ 580	6,350	4,874	6,985
Fuel Cell	$2,163						
2000		$2,143	$1,362	$2,176	5,787	5,760	6,000
2005		$2,028	$1,261	$2,176	5,574	5,521	6,000
2010		$1,626	$1,008	$2,176	5,361	5,281	6,000
2015		$1,460	$1,008	$2,176	5,361	5,281	6,000
2020		$1,383	$1,008	$2,176	5,361	5,281	6,000

Source: *AEO2000* National Energy Modeling System runs: aeo2k.d100199a, htecel.d100799a, ltecel.d100799a.

Table 44. Cost and Performance Characteristics for Renewable Energy Generating Technologies: Two Cases

| Technology | Base Costs Reference ($1998/kW) | Overnight Costs | | Capacity Factor | |
		Reference ($1998/kW)	High Renewables ($1998/kW)	Reference (%) Case	High Renewables (%) Case
Biomass	1,877				
2000		1,863	1,841	80	80
2005		1,748	1,681	80	80
2010		1,297	1,213	80	80
2015		1,148	1,044	80	80
2020		1,129	966	80	80
Geothermal	1,622				
2000		1,811	1,548	87	87
2005		1,393	1,398	87	87
2010		1,054	1,247	87	87
2015		1,045	1,095	87	87
2020		1,426	945	87	87
Wind	1,086				
2000		979	943	30	37
2005		961	857	32	41
2010		941	771	34	45
2015		917	685	36	46
2020		893	600	38	47
Solar Thermal	3,059				
2000		3,040	2,946	42	42
2005		2,894	2,730	42	44
2010		2,748	2,521	42	56
2015		2,602	2,319	42	68
2020		2,422	2,097	42	77
Photovoltaic	4,836				
2000		4,318	4,007	28	21
2005		2,434	3,294	28	21
2010		1,813	2,580	29	21
2015		1,719	1,867	30	21
2020		1,661	1,154	30	21

Source: *AEO2000* National Energy Modeling System runs: aeo2k.d100199a, hirenew.d100799a.

Notes and Sources

[87] Form EIA-861, Annual Electric Utility Report, 1997.

[88] Energy Information Administration, Integrating Module of the National Energy Modeling System: Model Documentation, DOE/EIA-M057(2000), (Washington, DC, December 1999).

[89] A registered utility holding company is defined as any company that owns or controls 10% of the voting securities of a public utility company. PUHCA defines a public utility company as any company that owns or operates generation, transmission, or distribution facilities for the sale of electricity to the public.

[90] Electric Power Research Institute and U.S. Department of Energy, Office of Utility Technologies *Renewable Energy Technology Characterizations* (EPRI TR-109496, December 1997) or http://www.eren.doe.gov/utilities/techchar.html.

Oil and Gas Supply Module

The NEMS Oil and Gas Supply Module (OGSM) constitutes a comprehensive framework with which to analyze oil and gas supply. A detailed description of the OGSM is provided in the EIA publication, *Model Documentation Report: The Oil and Gas Supply Module (OGSM)*, DOE/EIA-M063(2000), (Washington, DC, January 2000). The OGSM provides crude oil and natural gas short-term supply parameters to both the Natural Gas Transmission and Distribution Module and the Petroleum Market Module. The OGSM simulates the activity of numerous firms that produce oil and natural gas from domestic fields throughout the United States, acquire natural gas from foreign producers for resale in the United States, or sell U.S. gas to foreign consumers.

OGSM encompasses domestic crude oil and natural gas supply by both conventional and nonconventional recovery techniques. Nonconventional recovery includes enhanced oil recovery and unconventional gas recovery from tight gas formations, gas shale, and coalbeds. Foreign gas transactions may occur via either pipeline (Canada or Mexico) or transport ships as liquefied natural gas (LNG).

Primary inputs for the module are varied. One set of key assumptions concerns estimates of domestic technically recoverable oil and gas resources. Other major factors affecting the projection include the assumed rates of technological progress, the start date, and threshold price for the Alaskan Natural Gas Transportation System (ANGTS), projections for enhanced oil recovery production, supplemental gas supplies over time, and natural gas import and export capacities.

Key Assumptions

Domestic Oil and Gas Technically Recoverable Resources

Domestic oil and gas technically recoverable resources[91] consist of proved reserves,[92] inferred reserves,[93] and undiscovered technically recoverable resources.[94] OGSM resource assumptions are based on estimates of technically recoverable resources from the United States Geological Survey (USGS) and the Minerals Management Service (MMS) of the Department of the Interior. Supplemental adjustments to the USGS nonconventional resources are made by Advanced Resources International (ARI), an independent consulting firm, and to the deep resources in the Gulf of Mexico by the National Petroleum Council.[95] While undiscovered resources for Alaska are based on USGS estimates; estimates of recoverable resources are obtained on a field by field basis from a variety of sources including trade press. Published estimates in Tables 45 and 46 reflect the removal of intervening reserve additions between the dates of the USGS (1/1/94) and MMS (1/1/95) estimates and 1/1/98.

Alaskan Natural Gas

The outlook for natural gas production from the North Slope of Alaska is affected strongly by the unique circumstances regarding its transport to market. Unlike virtually all other identified deposits of natural gas in the United States, North Slope gas lacks a means of economic transport to major commercial markets. The lack of viable marketing potential at present has led to the use of Prudhoe Bay gas to maximize crude oil recovery in that field. This use is expected to delay extraction of gas for market until the post-2005 period. The estimates for gas from the North Slope that will be transported to lower 48 States markets through ANGTS are dependent on the capacity of this system. ANGTS is projected to flow gas to market in two phases, and it is assumed that production will be available to fully utilize the capacity in both phases, if constructed. Operational capacity for the first phase is 767 billion cubic feet per year delivered to the U.S./Canadian border. Annual capacity is assumed to increase to 1,150 billion cubic feet upon the completion of the second phase. Operation for each phase is assumed to begin at midyear; thus only half of the capacity is available for the first year of operation, with full capacity available in each year thereafter. It is assumed that ANGTS will not begin operation until 2005 at the earliest, to support oil recovery in the Prudhoe Bay field. Each phase of ANGTS is brought on line in OGSM when the appropriate border-crossing price is reached for gas delivered to the lower 48 States. The price for phase one is $4.00 in 1998 dollars per thousand cubic feet. When this price is reached, ANGTS is brought on line in the following year, with a total flow of 383 billion cubic feet, reaching the full capacity of 767 billion cubic feet in subsequent years. If a higher threshold price of $5.36, in 1998 dollars per thousand cubic feet is reached, then phase two will begin the following year. The flow will increase by 192 billion cubic feet, to 959 billion cubic feet, and in each subsequent year the flow will be 1,150 billion cubic feet. This methodology is applied in all the cases.

The projection for supplemental gas supply is identified for three separate categories: synthetic natural gas (SNG) from liquids, SNG from coal, and other supplemental supplies. SNG from the currently operating Great Plains Coal Gasification Plant is assumed to continue through 2008, at 1998 levels through 1999 and 55.68 billion cubic feet per year thereafter. In all cases, it is assumed that in midyear 2009 the Great Plains facility will stop producing natural gas when the current purchase contract expires and natural gas production is assumed not to be economical. Other supplemental supplies are held at a constant level of 49.14 billion cubic feet per year throughout the forecast because this level is consistent with historical data and there is no reason to believe this will change significantly in the context of a reference case forecast. Synthetic natural gas from liquid hydrocarbons is currently only produced in Hawaii. This production is assumed to continue over the forecast at the average historical level of 2.74 billion cubic feet per year.

Table 45. Crude Oil Technically Recoverable Resources
(Billion Barrels)

Crude Oil Resource Category	As of January 1, 1998
Undiscovered	57.21
Onshore	27.18
Offshore	30.03
Deep (>200 meter W.D.)	27.29
Shallow (0-200 meter W.D.)	2.74
Inferred Reserves	48.20
EOR	12.70
Other Onshore	31.40
Offshore	4.11
Deep (>200 meter W.D.)	1.87
Shallow (0-200 meter W.D.)	2.23
Total Lower 48 States Unproved	105.41
Alaska	10.53
Total U.S. Unproved	115.94
Proved Reserves	23.89
Total Crude Oil	139.83

Source: Energy Information Administration, Office of Integrated Analysis and Forecasting.

Note: Resources in restricted areas (where drilling is prohibited) are not included in this table. Also, the EOR and Alaska values are not explicitly utilized in the OGSM, but are included here to complete the table.

Table 46. Natural Gas Technically Recoverable Resources
(Trillion Cubic Feet)

Natural Gas Resource Category	As of January 1, 1998
Nonassociated Gas	
Undiscovered	355.83
Onshore	172.04
Deep (>10,000 ft)	86.82
Shallow (0-10,000 ft)	85.21
Offshore	183.79
Deep (>200 meters W.D.)	120.83
Shallow (0-200 meters W.D)	62.96
Inferred Reserves	222.82
Onshore	196.88
Deep (>10,000 ft)	22.67
Shallow (0-10,000 ft)	174.21
Offshore	25.93
Deep (>200 meters W.D.)	7.24
Shallow (0-200 (meters W.D.)	18.70
Unconventional Gas Recovery	377.85
• Tight Gas	270.54
• Shale	52.01
• Coalbed	55.31
Assc-Dissolved Gas	124.30
Total Lower 48 Unproved	1080.79
Alaska	11.46
Total U.S. Unproved	1092.25
Proved Reserves	167.22
Total Natural Gas	1259.47

Source: Energy Information Administration, Office of Integrated Analysis and Forecasting.

Note: Resources in restricted areas (where drilling is prohibited) are not included in this table. Also, the Associated Dissolved Gas and the Alaska values are not explicitly utilized in the OGSM, but are included here to complete the table.

Natural Gas Imports and Exports

U.S. natural gas trade with Mexico and natural gas exports from the United States to Canada are determined exogenously to NEMS. U. S. exports of LNG are also exogenously determined. Canadian production and U.S. import flows from Canada are determined endogenously within the model but are constrained by assumed pipeline capacities. Exogenously specified projections of pipeline import and export values from Canada and Mexico are shown in Table 47.

Table 47. U.S. Natural Gas Imports and Exports
(Billion Cubic Feet per Year)

| Year | Canada | | Mexico | |
	Import Capacity[1]	Exports	Imports	Exports
2000	4,598	57	36	70
2005	5,457	57	36	115
2010	5,578	57	36	170
2015	5,815	57	36	228
2020	6,141	57	36	235

[1]Canadian "import" figures represent design capacity, not actual flow projections, because flows are not an assumption. Canadian import flows are determined endogenously within the model.

Source: Energy Information Administration, Office of Integrated Analysis and Forecasting.

Note: Imports are imports to the United States. Exports are exports from the United States.

Canadian consumption and production outside of the Western Canadian Sedimentary Basin (WCSB) are set exogenously in the model and shown in Table 48. These values are reflective of a recent forecast produced by Canada's National Energy Board. Production in the WCSB is calculated endogenously to the model. In doing so, the natural gas finding rates are set across the forecast period by establishing an initial historical average finding rate of 1.57 billion cubic feet per well and assuming an annual decline of 2.5 percent.

Table 48. Exogenously Specified Canadian Production and Consumption
(Billion Cubic Feet per Year)

Year	Consumption	Production Eastern Canada	Production Northern Frontier
2000	2,711	170	0
2005	3,069	230	0
2010	3,345	395	0
2015	3,635	525	0
2020	3,908	555	45

Source: Canada's National Energy Board.

Annual U.S. exports of LNG were assumed to be a constant at 67.6 billion cubic feet in each projection year. LNG imports are determined endogenously within the model. The outlook for LNG imports was based on a combination of influences, including available gasification capacity, announced plans by each company, tanker availability, expected utilization rates, projected gas prices and liquefaction capacity, and long-term contracts with a responsible purchaser. LNG import capacity in 1998 is 359 billion cubic feet. The outlook for LNG imports also includes an implicit assumption that no major operational or institutional difficulties arise that are not resolved expeditiously.

Currently, only two LNG import terminals are in operation: the Distrigas facility in Everett, Massachusetts, and the Trunkline facility in Lake Charles, Louisiana. A third facility at Elba Island, Georgia, currently mothballed, is assumed to reopen in 2002, adding an additional 118 billion cubic feet capacity. The other existing import terminal, at Cove Point, Maryland, is not expected to reopen for tanker imports in the projection period.

Offshore Royalty Relief

The Outer Continental Shelf Deep Water Royalty Act (Public Law 104-58) gives the Secretary of Interior the authority to suspend royalty requirements on new production from qualifying leases and requires that royalty payments be waived on new leases sold in the 5 years following November 28, 1997. The volume of production on which no royalties are due is assumed to be 17.5 million barrels of oil equivalent (BOE) in water depths of 200 to 400 meters, 52.5 million BOE in water depths of 400 to 800 meters, and 87.5 million BOE in water depths greater than 800 meters. In any year during which the arithmetic average of the closing prices on the New York Mercantile Exchange for light sweet crude oil exceeds $28 per barrel or natural gas exceeds $3.50 per million Btu, any production of crude oil or natural gas will be subject to royalties at the lease stipulated royalty rate.

Climate Change Action Plan

The natural gas production forecasts incorporate the expected results of the Climate Change Action Plan (CCAP)— Action Item 35, entitled *Launch Coalbed Methane Outreach Program*. Under Action Item 35, the Department of Energy (DOE) and the Environmental Protection Agency (EPA) created a program to raise the awareness among key coal companies and State agencies of the potential for cost-effective methane emissions reduction.

Estimates of the production resulting from this program through 2020 have been obtained from EPA. These production projections are presented in Table 49.

Table 49. Production from Mines Reached by CCAP Action Item 35
(Billion cubic feet)

Year	Production
1997	13.9
1998	17.2
1999	20.4
2000	23.7
2005	26.4
2010	29.1
2015	31.8
2020	34.6

Source: Energy Information Administration, Office of Integrated Analysis and Forecasting.

The annual production increases resulting (linear interpolations for interim year) from CCAP Action Item 35 are added to baseline forecasts of coalbed methane (CBM) production from the OGSM. The additional production is allocated regionally based on sharing factors derived from analysis in the EPA report, *Opportunities to Reduce Anthropogenic Methane Emissions in the United States.*[96]

Rapid and Slow Technology Cases

Two alternative cases were created to assess the sensitivity of the projections to changes in the assumed rates of progress in oil and natural gas supply technologies. To create these cases a number of parameters representing technological penetration in the reference case were adjusted to reflect a more rapid and a slower penetration rate. In the reference case, the underlying assumption is that technology will continue to penetrate at historically observed rates. Since technologies are represented somewhat differently, in different submodules of the Oil and Gas Supply Module, the approach for representing rapid and slow technology penetration varied as well. For instance, the effects of technological progress on conventional oil and natural gas parameters in reference case such as, finding rates, drilling, lease equipment and operating costs, and success rates, were adjusted upward and downward by a third (Table 50), for the rapid and slow technology cases, respectively. The approaches taken in the representation of Canadian natural gas, enhanced oil recovery, and unconventional natural gas are discussed below. All other parameters in the model were kept at their reference case values, including technology parameters for other modules, parameters affecting foreign oil supply, and assumptions about imports and exports of LNG and natural gas trade between the United States and Mexico.

Canadian Gas

For consistency purposes, from 2000 forward Canadian consumption and key supply results were adjusted to stimulate assumed impacts of rapid and slow oil and gas technology penetration on Canadian markets. The exogenously specified Canadian consumption forecast was increased or decreased by a fraction (0.0020 for rapid, 0.0030 for slow) times the forecast year minus 1999, for the rapid and slow cases, respectively. As a result, by 2020 Canadian consumption was 4.2 percent higher in the rapid technology case and 6.3 percent lower in the slow technology case. On the supply side, the forecast for wells drilled and the associated average finding rate in the Western Canadian Sedimentary basin (WCSB) were adjusted as well. Using a similar structure as used in adjusting consumption, the forcasted wells were increased or decreased by 0.008 times the forcast year minus 1999, for the rapid and slow cases, respectively. However, since the forecasted wells are function of the wellhead price and the previous year's production level in Canada, the original valuse that are adjusted are already different in the technology cases from the reference case. Finally, the decline in the finding rate in the WCSB (set at 2.5 percent per year in the reference case) was set at 0.5 and 3.5 percent in the rapid and slow technology cases, respectively, from 2000 forward.

Table 50. Assumed Annual Rates of Technological Progress on Costs, Finding Rates, and Success Rates for Conventional Sources

(Percent)

Category	Natural Gas			Crude Oil		
	Slow Technological Progress	Reference	Rapid Technological Progress	Slow Technological Progress	Reference	Rapid Technological Progress
Costs						
Drilling						
• Onshore	0.86	1.29	1.72	0.86	1.29	1.72
• Offshore	1.35	2.02	2.69	1.35	2.02	2.69
• Alaska	0.67	1.00	1.33	0.67	1.00	1.33
Lease Equipment						
• Onshore	0.40	0.59	0.78	0.40	0.59	0.78
• Offshore	0.94	1.40	1.86	0.94	1.40	1.86
• Alaska	0.67	1.00	1.33	0.67	1.00	1.33
Operating						
• Onshore	0.36	0.54	0.72	0.36	0.54	0.72
• Offshore	0.40	0.60	0.80	0.40	0.60	0.80
• Alaska	0.67	1.00	1.33	0.67	1.00	1.33
Finding Rates						
New Field Wildcats						
Onshore						
• Shallow						
Northeast	0.34	0.50	0.67	0.34	0.50	0.67
Gulf Coast	1.34	2.00	2.66	1.34	2.00	2.66
Mid Continent	2.01	3.00	3.99	1.34	2.00	2.66
Southwest	2.01	3.00	3.99	2.68	4.00	5.32
Rocky Mountain	1.34	2.00	2.66	1.34	2.00	2.66
West Coast	0.67	1.00	1.33	0.67	1.00	1.33
• Deep						
Northeast	—	—	—	—	—	—
Gulf Coast	0.67	1.00	1.33	—	—	—
MidContinent	0.67	1.00	1.33	—	—	—
Southwest	4.02	6.00	7.98	—	—	—
Rocky Mountain	0.67	1.00	1.33	—	—	—
West Coast	—	—	—	—	—	—
Offshore	4.02	6.00	7.98	1.34	2.00	2.66
Other Exploratory						
Onshore						
• Shallow	0.00	0.00	0.00	1.93	2.88	3.83
• Deep	3.16	4.72	6.28	—	—	—
Offshore	2.77	4.14	5.51	2.77	4.14	5.51
Developmental						
Onshore						
• Shallow	0.18	0.27	0.36	1.68	2.50	3.33
• Deep	1.08	1.61	2.14	—	—	—
Offshore	2.77	4.14	5.51	2.77	4.14	5.51
Success Rate						
• Exploratory	0.34	0.50	0.67	0.34	0.50	0.67
• Developmental	0.00	0.00	0.00	0.00	0.00	0.00

Source: Energy Information Administration, Office of Integrated Analysis and Forecasting.

Enhanced Oil Recovery

Two impacts of technological improvements are modeled to determine the economics for development of inferred enhanced oil recovery (EOR) reserves: (1) an overall reduction in the costs of drilling, completion and equipping production wells due to incremental improvements in drilling equipment and procedures, reservoir characterization, completion methods, and operation refinement; and (2) the field-specific penetration of horizontal well technology, which represents a quantum improvement in recovery efficiency. The specific parameters for modeling the slow, reference, and rapid technology cases are shown in Table 51.

Table 51. Assumed Rates of Technological Progress on Enhanced Oil Recovery Techniques

Item	Slow Technology	Reference Technology	Rapid Technology
Decline in D,C,&E Costs (per year)	2%	2%	3%
Start Penetration of Horizontal Wells	NA	1995	1995
Horizontal Technology Penetration Period (years)	None	40	20
Horizontal Technology Penetration Rate (per year)	0%	2.5%	5%
Maximum Penetration of Inferred Reserve Base	0%	90%	90%

Source: Energy Information Administration, Office of Integrated Analysis and Forecasting.

The remaining undiscovered recoverable resource determined to be technically amenable to gas miscible EOR methods is set for each region at the beginning of the forecast assuming current technology. This

Table 52. Assumed Rates of Technological Progress for Gas Miscible EOR Methods

Region	Slow Technology	Reference Technology	Rapid Technology
2	0%	2.5%	3.5%
3	1%	2%	3%
4	1%	2%	3%
5	1%	2%	3%

Source: Energy Information Administration, Office of Integrated Analysis and Forecasting.

value is assumed to increase over the forecast period with advancements in technology (Table 52).

Unconventional Gas

The Unconventional Gas Recovery Supply Submodule (UGRSS) relies on the Technology Impacts and Timing functions to capture the effects of technological progress on costs and productivity in the development of gas from deposits of Coalbed Methane, Gas Shales, and Tight Sands. The numerous research and technology initiatives are combined into eleven specific "technology groups," that encompass the full spectrum of key disciplines — geology, engineering, operations and the environment. The technology groups utilized for the *Annual Energy Outlook 2000* are characterized for three distinct technology cases — Slow Technological Progress, Reference Case, and Rapid Technological Progress — that capture three different futures for technology progress. The eleven technology groups are presented below. Their treatment under the different technology cases are described in Table 53.

Unconventional Gas Recovery Technology Groups

1. <u>Basin Assessments</u>:   Basin assessments increase the available resource base by a.) accelerating the time that hypothetical plays in currently unassessed areas become available for development and b.) increasing the play probability for hypothetical plays - that portion of a given area that is likely to be productive.

2. Play Specific, Extended Reservoir Characterizations: Extended reservoir characterizations increase the pace of new development by accelerating the pace of development for emerging plays, where projects are assumed to require extra years for full development compared to plays currently under development..

3. Advanced Well Performance Diagnostics and Remediation: Well performance diagnostics and remediation expand the resource base by increasing reserve growth for already existing reserves.

4. Advanced Exploration and Natural Fracture Detection R&D: Exploration and natural fracture detection R&D increases the success of development by a.) improving exploration/development drilling success rates for all plays and b.) improving the ability to find the best prospects and areas.

5. Geology Technology Modelling and Matching: Geology/technology modelling and matching matches the "best available technology" to a given play with the result that the expected ultimate recovery (EUR) per well is increased.

6. More Effective, Lower Damage Well Completion and Stimulation Technology: Improved drilling and completion technology improves fracture length and conductivity, resulting in increased EUR's per well.

7. Targeted Drilling and Hydraulic Fracturing R&D: Targeted drilling and hydraulic fracturing R&D results in more efficient drilling and stimulation which lowers well drilling and stimulation costs.

8. New Practices and Technology for Gas and Water Treatment: New practices and technology for gas and water treatment result in more efficient gas separation and water disposal which lowers water and gas treatment operation and maintenance (O&M) costs.

9. Advanced Well Completion Technologies such as Cavitation, Horizontal Drilling, and Multi-lateral Wells: R&D in advanced well completion technologies a.) defines applicable plays, thereby accelerating the date such technologies are available and b.) introduces an improved version of the particular technology, which increases EUR per well.

10. Other Unconventional Gas Technologies, such as Enhanced Coalbed Methane and Enhanced Gas Shales Recovery: Other unconventional gas technologies introduce dramatically new recovery methods that a.) increase EUR per well and b.) become available at dates accelerated by increased R&D with c.) increased operation and maintenance (O&M) costs (in the case of Coalbed Methane) for the incremental gas produced.

11. Mitigation of Environmental Constraints: Environmental mitigation removes development constraints in environmentally sensitive basins, resulting in an increase in basin areas available for development.

Table 53. Assumed Rates of Technological Progress for Unconventional Gas Recovery

Technology Group	Item	Type of Deposit	Technology Case		
			Slow	Reference	Rapid
1	Year Hypothetical Plays Become Available	Coalbed Methane & Gas Shales	NA	2016	2013
		Gas Shales	NA	NA	2018
	Improvement in Play Probability for Hypothetical Plays (per year)	All Types	NA	NA	.5%
2	Decrease in Extended Portion of Development Schedule for Emerging Plays (per year)	Coalbed Methane & Gas Shales	NA	5.0%	7.5%
		Tight Sands	NA	6.3%	8.2%
3	Expansion of Existing Reserves (per year -declining .1% per year)	Coalbed Methane & Gas Shales	2.0%	3.0%	3.5%
		Tight Sands	1.0%	2.0%	2.5%
4	Increase in Percentage of Wells Drilled Successfully (per year)	Coalbed Methane & Gas Shales	.1%	.3%	.6%
		Tight Sands	.1%	.3%	.8%
	Year that Best 30 Percent of Basin is Fully Identified	Coalbed Methane	2017	2017	2012
		Tight Sands	NA	2017	2012
		Gas Shales	NA	2017	2017
5	Increase in EUR per Well (per year)	Coalbed Methane & Gas Shales	NA	.1%	.3%
		Tight Sands	NA	.3%	.4%
6	Increase in EUR per Well (per year)	Coalbed Methane & Gas Shales	.3%	.4%	.6%
		Tight Sands	.4%	.5%	.6%
7	Decrease in Drilling and Stimulation Costs per Well (per year)	Coalbed Methane & Gas Shales	.3%	.5%	.8%
		Tight Sands	.3%	.5%	.5%
8	Decrease in Water and Gas Treatment O&M Costs per Well (per year)	All Types	.6%	1.0%	1.3%
9	Year Advanced Well Completion Technologies Become Available	Coalbed Methane	2018	2011	2008
		Tight Sands	2016	2011	2008
		Gas Shales	NA	NA	2016
	Increase in EUR per well (total increase)	Coalbed Methane	10%	20%	25%
		Tight Sands	5%	10%	12.5%
		Gas Shales	NA	NA	5%
10	Year Advanced Recovery Technologies Become Available	Coalbed Methane	2018	2010	2010
		Tight Sands	NA	NA	2018
		Gas Shales	NA	NA	NA
	Increase in EUR per well (total increase)	Coalbed Methane	10%	25%	27.5%
		Tight Sands	NA	NA	10%
		Gas Shales	NA	NA	NA
	Increase in Costs ($1998/Mcf) for Incremental CBM production	Coalbed Methane	1.54	1.03	.91
		Tight Sands & Gas Shales	NA	NA	NA
11	Proportion of Areas Currently Restricted that Become Available for Development (per year)	Coalbed Methane	.3%	1%	1.3%
		Tight Sands	NA	1%	1.3%
		Gas Shales	.5%	1%	1.3%

EUR = Expected Ultimate Recovery.

O&M = Operation & Maintenance.

CBM = Coalbed Methane.

Source: Energy Information Administration, Office of Integrated Analysis and Forecasting.

Notes and Sources

[91] *Economically recoverable resources* are those volumes considered to be of sufficient size and quality for their production to be commercially profitable by current conventional or nonconventional technologies, under specified economic conditions.

[92] *Proved reserves* are the estimated quantities that analysis of geological and engineering data demonstrate with reasonable certainty to be recoverable in future years from known reservoirs under existing economic and operating conditions.

[93] *Inferred reserves* are that part of expected ultimate recovery from known fields in excess of cumulative production plus current reserves.

[94] *Undiscovered resources* are located outside oil and gas fields in which the presence of resources has been confirmed by exploratory drilling; they include resources from undiscovered pools within confirmed fields when they occur as unrelated accumulations controlled by distinctly separate structural features or stratigraphic conditions.

[95] Donald L. Goutier and others, U.S. Department of Interior, U.S. Geological Survey, *1995 National Assessment of the United States Oil and Gas Resources*, (Washington, D.C., 1995); U.S. Department of Interior, Minerals Management Service, an Assessment of the Undiscovered Hydrocarbon Potential of the Nation's Outer Continental Shelf, OGS Report MMS 96-0034 (June 1996).

[96] United States Environmental Protection Agency, *Opportunities to Reduce Anthropogenic Emissions in the United States: Report to Congress,* EPA430-R-93-012, (Washington, DC, October 1993).

Natural Gas Transmission and Distribution Module

The NEMS Natural Gas Transmission and Distribution Module (NGTDM) derives domestic natural gas production, wellhead and border prices, end-use prices, and flows of natural gas through the regional interstate network, for both a peak (December through March) and off peak period during each forecast year. These are derived by solving for the market equilibrium across the three main components of the natural gas market: the supply component, the demand component, and the transmission and distribution network that links them. In addition, natural gas flow patterns are a function of the pattern in the previous year, coupled with the relative prices of gas supply options as translated to the represented market "hubs." The major assumptions used within the NGTDM are grouped into five general categories. They relate to (1) the classification of demand into core and noncore transportation service classes, (2) the pricing of transmission and distribution services, (3) pipeline and storage capacity expansion and utilization, (4) the implementation of recent regulatory reform, and (5) the implementation of provisions of the Climate Change Action Plan (CCAP). A complete listing of NGTDM assumptions and in-depth methodology descriptions are presented in *Model Documentation: Natural Gas Transmission and Distribution Model of the National Energy Modeling System, Model Documentation 2000, DOE/EIA-M062(2000), January 2000.*

Key Assumptions

Demand Classification

Customers demanding natural gas are classified as either core or noncore customers, with core customers assumed to transport their gas under firm (or near firm) transportation agreements and noncore customers assumed to transport their gas under interruptible or short-term capacity release transportation agreements. A distinction is made between core and noncore customers because the price differentials can be significant and it allows for a different algorithm to be used in setting the prices. All residential, commercial, and transportation (vehicles using compressed natural gas) end-use customers are assumed to be core customers. Industrial customers fall into both categories, with industrial boilers and refineries assumed to be noncore and all other industrial users assumed to be core. Likewise, customers in the electric generator sector are assumed to be both core and noncore.[97] Gas steam and gas combined-cycle units are considered to be core; and the remaining units are classified as noncore.

End-use sector specific load patterns are based on recent historical patterns and do not change over the forecast, with the exception of the electric generation sector[98] (i.e., there is no representation of changes in load patterns from new technologies like natural gas cooling.) However, pipeline load factors do change over the forecast as the composition of end-use changes across sectors and as more pipeline and storage capacity becomes available.

Pricing of Services

Transportation rates for interstate pipeline services (both between NGTDM regions and within a region) are calculated assuming that the costs of new pipeline capacity will be rolled into the existing rate base. The flow of gas in the peak period is based on reservation and usage charges; while the off-peak flows are just based on usage fees. While cost-of-service still forms the basis for pricing these services, an adjustment to the tariffs is made based on changes in utilization to reflect a more market-based approach. Capital expenditures for refurbishment are generally relatively small, are offset by retirements, and are therefore not considered, nor are potential future expenditures for pipeline safety (refurbishment costs include any expenditures for repair and/or replacement of existing pipe). Existing gross plant in service is only based on new capacity addition.

End-use prices for residential, commercial, and core industrial customers are derived by adding a markup to the regional hub price of natural gas in both peak and off-peak periods. (Prices are only reported on an annual basis and represent quantity-weighted averages of the two seasons.) These markups include the cost of service provided by intraregional interstate pipelines, intrastate pipelines, and local distributors. The intrastate tariffs are accounted for endogenously through historical model benchmarking. Distributor tariffs represent the difference between the regional end-use and citygate price, independent of whether or not a customer class typically purchase gas through a local distributor. The distribution tariffs are initially based on 1998 historical data (Table 54), but they are adjusted throughout the forecast in response to changes in consumption levels and cost of labor and capital. In addition, a decline rate of 1 percent per year is applied (independent of changes in costs related to the cost of capital and labor and consumption levels) to account for capital depreciation combined with efficiency improvements. Although the markups in Table 54 represent annual averages, the model actually uses separate markups for the peak and offpeak periods.

End-use prices for noncore industrial and electric generator customers are established by adding a markup to the natural gas market price at the corresponding core or noncore segment at the regional market hub. These markups are endogenously derived as the difference between estimated historical 1998 end-use

Table 54. Base Year Average 1998 Annual Distributor Markup for Local Transportation Service
(1998 Dollars per Thousand Cubic Feet)

Region	Residential	Commercial	Core Industrial	Core Electric Generators
New England	5.44	3.01	-0.25	-0.92
Mid Atlantic	4.96	1.75	0.57	-0.51
East North Central	2.52	2.03	0.05	-0.93
West North Central	2.75	1.68	-0.12	-0.77
South Atlantic	5.28	3.57	0.39	-0.59
East South Central	3.39	2.41	-0.35	-0.54
West South Central	3.60	2.03	0.26	-0.27
Mountain	2.77	1.89	0.46	-0.30
Pacific	3.96	2.05	4.01	-0.27
Florida	8.15	3.23	-1.09	-0.78
Arizona/New Mexico	4.55	2.68	0.47	0.03
California	4.54	3.82	0.96	0.30

Source: Energy Information Administration, Office of Integrated Analysis and Forecasting. Derived from Form EI-857, "Monthly Report of Natural Gas Purchases and Deliveries to Consumers" for residential, commercial, and citygate, from the *Manufacturing Energy Consumption Survey Consumption of Energy 1994*, (Form EIA-846) for core industrial, and derived from Form FERC-423 for core electric generators, *Monthly Report of Cost and Quality of Fuels for Electric Plants* for core electric.

prices[99], and the NGTDM regional hub price, and held constant throughout the forecast. End-use prices for core electric generator customers are similarly established with markups initially based on 1998 end-use prices. However, these markups are adjusted each forecast year by a fraction (0.05) of the annual percentage change in the core electric generator consumption. This adjustment is intended to reflect anticipated additional infrastructure devoted to serving core electric generation consumption growth.

The vehicle natural gas (VNG) sector is divided into fleet and non-fleet vehicles. The distributor tariffs for natural gas to fleet vehicles are set to *EIA's Natural Gas Annual* historical end-use minus citygate prices plus Federal and State VNG taxes (Table 55). The price to non-fleet vehicles is based on the industrial sector firm price plus an assumed $4.07 (1998 dollars per thousand cubic feet) dispensing charge plus Federal and State taxes, set constant in nominal dollars. It is assumed that the retailer will lower the dispensing charge by up to 20 percent if needed to be competitive with gasoline prices.

Table 55. Vehicle Natural Gas (VNG) Pricing
(Nominal dollars per thousand cubic feet)

Modified Census Divisions	Total Federal and State VNG Tax[1]
New England	2.16
Middle Atlantic	2.52
East North Central	1.81
West North Central	1.53
South Atlantic (excludes Florida)	1.81
East South Central	0.75
West South Central	1.60
Mountain (excludes Arizona and New Mexico)	0.85
Pacific (excludes California)	2.40
Florida	1.13
Arizona and New Mexico	0.27
California	0.70

[1]Assuming a $0.4844 (nominal dollars per thousand cubic feet) Federal Tax.

Source: Energy Information Administration, Office of Integrated Analysis and Forecasting, based on the Federal tax published in the Information Resources, Inc., publication *Octane Week*, August 9, 1993, and State taxes posted at the Department of Energy website titled "Alternative Fuels Data Center" at www.afdc.doe.gov.

Capacity Expansion and Utilization

For the first 2 forecast years of the model, announced pipeline and storage capacity expansions (that are deemed highly likely to occur) are used to establish limits on flows and storage in the model. Subsequently, pipeline and storage capacity is added when increases in demand, coupled with anticipated price impacts, warrant such additions (i.e., flow is allowed to exceed current capacity if the demand still exists given the adjusted tariff, thus indicating an expansion). When the decision to add capacity is made, a simple representation is incorporated to capture the average capital costs for pipeline and storage expansion and the resulting tariff. Once it is determined that an expansion will occur, the associated capital costs are estimated based on costs of recent expansions in that area and are used in the revenue requirement calculations in future years.

It is assumed that pipelines and local distribution companies build and subscribe to a portfolio of pipeline and storage capacity to serve a region-specific colder-than-normal winter demand level, currently set at 5 percent for all pipeline area. Maximum pipeline capacity utilization in the peak period is set at 99 percent. In the off-peak period, the maximum is assumed to vary between 75 and 99 percent of the design capacity. The overall level and profile of consumption as well as the availability and price of supplies generally cause realized pipeline utilization levels to be lower than the maximum. For each sector, consumption is disaggregated into peak and off-peak periods based on average historical patterns. In current form, time of use pricing can not be modeled.

Additions to underground storage capacity are constrained to capture limitations of geology in each of the market regions. The constraints limit total storage additions to be less than an expansion factor times the 1990 storage capacity.

The model methodology represents net injections of natural gas into storage in the off-peak period and net withdrawals during the peak period. Total annual net storage withdrawals equal zero in all years of the forecast, which would be expected under normal weather conditions.

Legislation and Regulation

The methodology for setting reservation fees for transportation services is consistent with FERC's alternative ratemaking and capacity release position in that it allows flexibility in the rates pipelines charge. The methodology is market-based in that prices for transportation services will respond positively to increased demand for services while prices will decline (reflecting discounts to retain customers) should the demand for services decline. The model also reflects current legislation and regulation.

Climate Change Action Plan

The Climate Change Action Plan (CCAP) initiatives to increase the natural gas share of total energy use through Federal regulatory reform (Action 23) are reflected in the methodology for the pricing of pipeline services. Provisions of the CCAP to expand the Natural Gas Star program (Action 32) are assumed to recover 35 billion cubic feet of natural gas per year from 2000 through the end of the forecast period that otherwise might be lost to fugitive emissions.

Notes and Sources

[97]   The electric generator end-use category includes gas consumption by any facility whose sole purpose is electricity generation (including independent power producers). Natural gas consumption by cogenerators (producers of electricity as a by-product of another process) is included in industrial end-use consumption.

[98]   Natural gas consumption by electric generators is established in the Electricity Market Module of NEMS on a seasonal basis. These values are used as a basis for adjusting the related load patterns throughout the forecast.

[99]   Historical core and noncore industrial prices were based on data from the *Manufacturing Consumption of Energy 1991*, 1994.

Petroleum Market Module

The NEMS Petroleum Market Module (PMM) forecasts petroleum product prices and sources of supply for meeting petroleum product demand. The sources of supply include crude oil (both domestic and imported), petroleum product imports, other refinery inputs including alcohol and ethers, natural gas plant liquids production, and refinery processing gain. In addition, the PMM estimates capacity expansion and fuel consumption of domestic refineries.

The PMM contains a linear programming representation of refining activities in three U.S. regions. This representation provides the marginal costs of production for a number of traditional and new petroleum products. The linear programming results are used to determine end-use product prices for each Census Division using the assumptions and methods described below.[100]

Key Assumptions

Product Types and Specifications

The PMM models refinery production of the products shown in Table 56.

The costs of producing new formulations of gasoline and diesel fuel that will be phased in as a result of the Clean Air Act Amendments of 1990 (CAAA90) are determined within the linear programming representation by incorporating specifications and demands for these fuels. The PMM assumes that the specifications for these new fuels will remain the same as specified in current legislation.

Table 56. Petroleum Product Categories

Product Category	Specific Products
Motor Gasoline	Traditional Unleaded, Oxygenated, Reformulated
Jet Fuel	Kerosene-type
Distillates	Kerosene, Heating Oil, Highway Diesel
Residual Fuels	Low Sulfur, High Sulfur
Liquefied Petroleum Gases	Propane, Liquefied Petroleum Gases Mixed
Petrochemical Feedstocks	Petrochemical Naptha, Petrochemical Gas Oil, Propylene, Aromatics
Others	Lubricating products and Waxes, Asphalt/Road Oil, Still Gas Petroleum Coke, Special Naphthas

Source: Energy Information Administration, Office of Integrated Analysis and Forecasting.

The PMM models the production and distribution of three different types of gasoline: traditional, oxygenated, and reformulated phase 2. The following specifications are included in PMM to differentiate between traditional and reformulated gasoline blends (Table 57): oxygen content, Reid vapor pressure (Rvp), benzene content, aromatic content, sulfur content, olefin content, and the percent evaporated at 200 and 300 degrees Fahrenheit (E200 and E300).

Traditional gasoline must comply with antidumping requirements aimed at preventing the quality of traditional gasoline from eroding as the reformulated gasoline program is implemented. Traditional gasoline must meet the Complex Model compliance standards which cannot exceed average 1990 levels of toxic and nitrogen oxide emissions.[101]

Oxygenated gasoline, which has been required during winter in many U.S. cities since October of 1992, requires an oxygenated content of 2.7 percent by weight. Oxygenated gasoline is assumed to have specifications identical to traditional gasoline with the exception of a higher oxygen requirement. Some areas that require oxygenated gasoline will also require reformulated gasoline. For the sake of simplicity, the areas of overlap are assumed to require gasoline meeting the reformulated specifications.

Table 57. Year Round Gasoline Specifications by Petroleum Administration for Defense Districts (PADD)

PADD	Reid Vapor Pressure (Max)	Oxygen Weight Percent (Min)	Oxygen Weight Percent (Max)	Aromatics Volume Percent (Max)	Benzene Volume Percent (Max)	Sulfur PPM (Max)	Olefin Volume Percent (Max)	Percent Evaporated at 200°	Percent Evaluated at 300°
Traditional									
PADD I-V	10.0	—	—	28.6	1.6	338.4	10.8	41.0	83.0
PADD V	9.2	—	—	28.6	1.6	338.4	10.8	41.0	83.0
Reformulated									
PADD I-IV	8.5	2.1	2.7	25.0	0.95	135.0	12.0	49.0	87.0
PADD V	7.9	1.7	1.8	25.0	0.72	25.0	6.0	49.0	85.0

Max = Maximum.

Min = Minimum.

PADD = Petroleum Administration for Defense District.

PPM = Parts per million by weight.

Source: Energy Information Administration, Office of Integrated Analysis and Forecasting.

Reformulated gasoline has been required in many areas in the U.S. since January 1995 (Table 58). In 1998, the EPA began certifying reformulated gasoline using the "complex model," which allows refiners to specify reformulated gasoline based on emissions reductions from their company, 1990 baseline or the EPA's 1990 baseline. The PMM reflects "Phase II" reformulated gasoline requirements which begin in 2000. The PMM uses a set of specifications that meet the "complex model" requirements, but it does not attempt to determine the optimal specifications that meet the "complex model." (Table 57).

The CAAA90 provided for special treatment of California that would allow different specifications for oxygenated and reformulated gasoline in that State. In 1992, California requested a waiver from the winter oxygen requirements of 2.7 percent to reduce the requirement to a range of 1.8 to 2.2 percent. The PMM assumes that Petroleum Administration for Defense District (PADD) V refiners must meet the California specifications. The specifications for reformulated gasoline in PADD V are the same as California standards.

In March 1999 California Governor Gray Davis issued an executive order announcing that the use of the gasoline blending agent, methyl tertiary butyl ether (MTBE) will be banned in the Sates by the end of 2002, due to water contamination problems. *AEO200* reflects this ban on MTBE in gasoline. Californian's congressmen have proposed legislation that would waive the Federal requirement for oxygen in areas covered by the Federal reformulated gasoline program (Los Angeles, San Diege and Sacramento). *AEO2000* assumes that the Federal oxygen requirement remains intact in these areas because no waiver had been granted at the time of production.

Rvp limitations are effective during summer months, which are defined differently in different regions. In addition, different Rvp specifications apply within each refining region, or PADD. The PMM assumes that these variations in Rvp are captured in the annual average specifications, which are based on summertime Rvp limits, wintertime estimates, and seasonal weights.

Motor Gasoline Market Shares

Within the PMM, total gasoline demand is disaggregated into demand for traditional, oxygenated, and reformulated gasoline by applying assumptions about the annual market shares for each type. The shares are able to change over time based on assumptions about the market penetration of new fuels. In *AEO2000*, the annual market shares for each region reflect actual 1998 market shares and are held constant throughout the forecast. The Census Divisions 3 and 4 market shares were adjusted because St. Louis,

Missouri, joined the Federal reformulated gasoline program in the summer of 1999. (See Table 58 for *AEO2000* market share assumptions.)

Table 58. Market Share for Gasoline Types by Census Division
(Percentage)

Gasoline Type/Year	Census Division								
	1	2	3	4	5	6	7	8	9
Traditional Gasoline	19	42	81	69	82	95	72	73	21
Oxygenated Gasoline (2.7% oxygen)	0	0	0	20	0	0	2	14	5
Reformulated Gasoline (2.0% oxygen)	81	58	19	11	18	5	27	13	74*

Source: Energy Information Administration, Office of Integrated Analysis and Forecasting.

*Note: 46 percent is assumed to continue the 2.0 percent Federal oxygen requirement. 28 percent is not covered by this requirement.

Diesel Fuel Specifications and Market Shares

In order to account for diesel desulfurization regulations related to CAAA90, low-sulfur diesel is differentiated from other distillates. In NEMS diesel fuel in Census Divisions 1 through 8 is required to meet Federal requirements, while diesel fuel in Census Division 9 is required to meet California Air Resources Board (CARB) standards. Both Federal and CARB standards limit sulfur to 500ppm.

The PMM contains a sharing methodology to allocate distillate demands between low and high sulfur. Market shares for low-sulfur diesel and distillate fuel are estimated based on data from EIA's annual *Fuel Oil and Kerosene Sales 1997*, (on line: http://www.eia.doe.gov/oil_gas/fok/1996/fokframe96.html, November 3, 1997). Since about 17 percent of current demand in the transportation sector is off highway, 83 percent of transportation demand for distillate fuel is assumed to be low sulfur. Consumption of low-sulfur distillate also occurs in the industrial sector where it is 36 percent of the market.

End-Use Product Prices

End-use petroleum product prices are based on marginal costs of production plus production-related fixed costs plus distribution costs and taxes. The marginal costs of production are determined by the model and represent variable costs of production including additional costs for meeting reformulated fuels provisions of the CAAA90. Environmental costs associated with controlling pollution at refineries. (Table 59) are reflected as fixed costs (associated operation and maintenance costs prior to 1996 are excluded).[102] Assuming that refinery-related fixed costs are recovered in the prices of light products, fixed costs are allocated among the prices of liquefied petroleum gases, gasoline, distillate, kerosene, and jet fuel. These costs are based on average annual estimates and are assumed to remain constant over the forecast period.

Table 59. Summary of Refinery Site Environmental Costs by Petroleum Administration for Defense Districts (PADD)
(1998 Dollars per Barrel)

Cost Category	PADD I	PADD II	PADD III	PADD IV	PADD V
Environmental Costs	0.65	0.66	0.52	0.96	0.73

PADD = Petroleum Administration for Defense District.

Source: Energy Information Administration, Office of Integrated Analysis and Forecasting.

The costs of distributing and marketing petroleum products are represented by adding fixed distribution costs to the marginal and refinery fixed costs of products. The distribution costs are applied at the Census Division level (Table 60) and are assumed to be constant throughout the forecast and across scenarios.

Table 60. Petroleum Product End-Use Markups by Sector and Census Division
(1998 Dollars per Gallon)

Sector/Product	Census Division								
	1	2	3	4	5	6	7	8	9
Residential Sector									
Distillate Fuel Oil	0.37	0.44	0.31	0.26	0.42	0.30	0.20	0.29	0.38
Kerosene	0.52	0.58	0.47	0.41	0.51	0.32	0.47	0.59	0.89
Liquefied Petroleum Gases	0.89	0.93	0.51	0.35	0.77	0.65	0.57	0.54	0.81
Commercial Sector									
Distillate Fuel Oil	0.14	0.12	0.05	0.03	0.06	0.04	0.04	0.04	0.07
Gasoline	0.14	0.13	0.13	0.13	0.13	0.17	0.17	0.16	0.19
Kerosene	0.27	0.22	0.20	0.14	0.19	0.25	0.25	0.19	0.23
Liquefied Petroleum Gases	0.53	0.54	0.44	0.34	0.53	0.40	0.30	0.43	0.55
Low-Sulfur Residual Fuel Oil	0.02	0.06	0.03	0.00	0.04	0.03	-0.01	0.00	0.08
Utility Sector									
Distillate Fuel Oil	0.02	0.02	0.02	0.01	0.01	0.07	0.03	0.05	0.01
High-Sulfur Residual Fuel Oil[3]	0.01	0.02	0.10	-0.04	0.00	-0.05	0.08	0.01	0.08
Low-Sulfur Residual Fuel Oil[3]	-0.01	0.00	0.09	-0.06	0.01	-0.06	0.06	0.21	0.17
Transportation Sector									
Distillate Fuel Oil	0.23	0.17	0.14	0.11	0.14	0.13	0.13	0.14	0.19
E85[1]	0.26	0.26	0.26	0.26	0.26	0.26	0.26	0.26	0.26
Gasoline	0.14	0.12	0.13	0.15	0.13	0.17	0.17	0.16	0.13
High-Sulfur Residual Fuel Oil[3]	-0.01	0.03	0.13	-0.04	-0.01	-0.07	0.05	0.20	0.10
Jet Fuel[4]	-0.01	-0.01	-0.02	-0.04	-0.04	-0.01	0.00	-0.01	0.00
Liquefied Petroleum Gases	0.49	0.48	0.48	0.32	0.44	0.34	0.26	0.35	0.48
M85[2]	0.14	0.12	0.13	0.15	0.13	0.17	0.18	0.17	0.13
Industrial Sector									
Asphalt and Road Oil	0.21	0.12	0.25	0.27	0.14	0.09	0.19	0.30	0.16
Distillate Fuel Oil	0.15	0.14	0.11	0.10	0.10	0.09	0.09	0.09	0.13
Gasoline	0.15	0.13	0.14	0.15	0.13	0.17	0.17	0.16	0.13
High-Sulfur Residual Fuel Oil	0.00	0.00	0.00	0.00	0.00	0.00	0.00	0.00	0.00
Kerosene	0.27	0.22	0.21	0.14	0.18	0.25	0.24	0.22	0.23
Liquefied Petroleum Gases	0.46	0.45	0.45	0.26	0.45	0.28	0.13	0.22	0.50
Low-Sulfur Residual Fuel Oil	0.01	0.03	0.04	0.01	0.03	0.03	0.04	0.09	0.09

[1] 85 percent ethanol and 15 percent gasoline.

[2] 85 percent methanol and 15 percent gasoline.

[3] Negative values indicate that an average end-use sales prices were less than wholesale prices. This often occurs with residual fuel which is produced as a biproduct when crude oil is refined to make higher value products like gasoline and heating oil.

[4] Negative values indicate that an average end-use sales prices were less than wholesale prices. This often occurs with jet fuel due to discounted sales to large volume end-users (airlines).

Note: Use conversion factors listed in Table H1 of the *Annual Energy Outlook 2000* to convert values to physical units.

Sources: Markups based on data from Energy Information Administration (EIA), Form EIA-782A, *Refiners'/Gas Plant Operators' Monthly Petroleum Product Sales Report*; EIA, Form EIA-782B, *Resellers'/Retailers' Monthly Petroleum Report Product Sales Report*; EIA, Form FERC-423, *Monthly Report of Cost and Quality of Fuels for Electric Plants*; EIA, Form EIA-759 *Monthly Power Plant Report*; EIA, *State Energy Data Report 1996*, DOE/EIA-0214(96), (Washington, DC, February 1999); EIA, *State Energy Price and Expenditures Report 1995*, DOE/EIA-0376(95), (Washington, DC, August 1998); and EIA, *Petroleum Marketing Monthly March 1999*, DOE/EIA-0380(99/03), (Washington, DC, March 1999).

Distribution costs for each product, sector, and Census Division represent average historical differences between end-use and wholesale prices. The distribution costs for kerosene are the average difference between end-use prices of kerosene and wholesale distillate prices. Distribution costs for M85 are assumed to be equivalent to distribution costs for gasoline.

State and Federal taxes are also added to transportation fuels to determine final end-use prices (Tables 61 and 62). Recent tax trend analysis indicated that State taxes increase at the rate of inflation, therefore, State taxes are held constant in real terms throughout the forecast. This assumption is extended to local taxes which are assumed to average 2 cents per gallon.[103] Federal taxes are assumed to remain at current levels in accordance with the overall *AEO2000* assumption of current laws and regulation. Federal taxes are deflated as follows:

$$\text{Federal Tax}_{product, year} = \text{Current Federal Tax}_{product} / \text{GDP Deflator}_{year}$$

Table 61. State and Local Taxes on Petroleum Transportation Fuels by Census Division
(1998 Dollars per Gallon)

| Year/Product | Census Division | | | | | | | | |
	1	2	3	4	5	6	7	8	9
Gasoline[1]	0.26	0.22	0.24	0.21	0.18	0.20	0.21	0.23	0.24
Diesel	0.20	0.26	0.25	0.20	0.17	0.16	0.19	0.21	0.22
Liquefied Petroleum Gases	0.11	0.13	0.16	0.17	0.16	0.16	0.15	0.09	0.05
M85[2]	0.25	0.18	0.19	0.14	0.13	0.16	0.19	0.20	0.12
E85[3]	0.25	0.18	0.16	0.19	0.13	0.16	0.19	0.15	0.12
Jet Fuel	0.04	0.02	0.01	0.03	0.04	0.03	0.00	0.03	0.02

[1]Tax also applies to gasoline consumed in the commercial and industrial sectors.

[2] 85 percent methanol and 15 percent gasoline.

[3] 85 percent ethanol and 15 percent gasoline.

Source: Aggregated from Federal Highway Administration, Tax Rates on Motor Fuel February 1, 1999, Table MF-121T, http://www.fhwa.dot.gov/ohim/novmmfr.pdf, (Washington, DC, March 1999). *Clean Fuels Report* (Washington, DC, February 1999).

Table 62. Federal Taxes
(Nominal Dollars per Gallon)

Product	Tax
Gasoline	0.18
Diesel	0.24
Jet Fuel	0.04
Liquefied Petroleum Gases	0.14
M85[1]	0.09
E85[2]	0.13

[1] 85 percent methanol and 15 percent gasoline.

[2] 85 percent ethanol and 15 percent gasoline

Sources: Omnibus Budget Reconciliation Act of 1993 (H.R. 2264); Tax Payer Relief Act of 1997 (PL 105-34) and *Clean Fuels Report* (Washington, DC, April 1998).

Crude Oil Quality

In the PMM, the quality of crude oil is characterized by average gravity and sulfur levels. Both domestic and imported crude oil are divided into five categories as defined by the ranges of gravity and sulfur shown in Table 63.

A "composite" crude oil with the appropriate yields and qualities is developed for each category by averaging the characteristics of specific crude oil streams that fall into each category. While the domestic and foreign categories are the same, the composite crudes for each category may differ because different crude streams make up the composites. For domestic crude oil, estimates of total regional production are made first, then shared out to each of the five categories based on historical data. For imported crude oil, a separate supply

Table 63. Crude Oil Specifications

Crude Oil Categories	Sulfur (percent)	Gravity (degrees API)
Low Sulfur Light	0 - 0.5	> 24
Medium Sulfur Heavy	0.35 - 1.1	> 24
High Sulfur Light	> 1.1	> 32
High Sulfur Heavy	> 1.1	24 - 33
High Sulfur Very Heavy	> 0.7	0 - 23

Source: Energy Information Administration, Office of Integrated Analysis and Forecasting.

curve is provided for each of the five categories.

Regional Assumptions

PMM reflects three refining regions: PADD I, PADD V, and a third region including PADD II-IV. Individual refineries are aggregated into one linear programming representation for each region. In order to interact with other NEMS modules with different regional representations, certain PMM inputs and outputs are converted from a PMM region to a non-PMM regional structure and vice versa.

Cogeneration Assumptions

Electricity consumption in the refinery is a function of the throughput of each unit. Sources of electricity consist of refinery power generation, utility purchases, refinery cogeneration, and merchant cogeneration. Power generators and cogenerators are modeled in the PMM LP as separate units which are allowed to compete along with purchased electricity. Both the refinery and merchant cogeneration units provide estimates of capacity, fuel consumption, and electricity sales to grid based on historical parameters.

Refinery sales to the grid are estimated using the following perecentages which are based on 1997 data:

Region	Percent Sold To Grid
1 (PADD I)	55.8
2 (PADD's II, III, and IV	3.8
3 (PADD V)	19.9

The PMM is forced to sell electricity back to the grid in these percentages at a price equal to the average price of electricity.

Merchant cogenerator's are defined as non-refiner owned facilities located near refineries to provide energy to the open market and to the neighboring refinery. The PMM assumes that 67 percent of electricity from merchant cogenerators in every region is sold to the grid. These sales occur at a price equal to the average of the generation price and the industrial price of electricity for each PMM region. Electricity prices are obtained from the Electricity Market Model.

Capacity Expansion Assumptions

PMM allows for capacity expansion of all processing units including distillation capacity, vacuum distillation, hydrotreating, coking, fluid catalytic cracking, hydrocracking, alkylation, and methyl tertiary butyl ether manufacture. Capacity expansion occurs by processing unit, starting from base year capacities established by PADD using historical data.

Expansion occurs in NEMS when the value received from the additional product sales exceeds the investment and operating costs of the new unit. The investment costs assume a 15-percent hurdle rate in the decision to invest and a 15-percent rate of return over a 15-year plant life. Expansion through 1999 is determined by adding to the existing capacities of units planned and under construction that are expected to begin operating during this time. Capacity expansion plans are done every 3 years. For example, after the model has reached a solution for forecast year 2001, the PMM looks ahead and determines the optimal capacities given the demands and prices existing in the 2004 forecast year. The PMM then allows 50 percent of that capacity to be built in forecast year 2002, 25 percent in 2003, and 25 percent in 2004. At the end of 2004, the cycle begins anew, looking ahead to 2007.

Strategic Petroleum Reserve Fill Rate

AEO2000 assumes no additions for the Strategic Petroleum Reserve during the forecast period. Additions to the Strategic Petroleum Reserve have not been included in recent budgets.

Short-term Methodology

Petroleum balance and price information for the year 1999 is projected at the U.S. level in the *Short-term Energy Outlook, September 1999* (*STEO*). The PMM assumes the STEO results for 1999, using regional estimates derived from the national STEO projections.

Biofuels (Ethanol) Supply

Background

The PMM provides supply functions on an annual basis through 2020 for ethanol produced from both corn and cellulosic biomass to produce transportation fuel.

Assumptions

- Corn feedstock supplies and costs are provided exogenously to NEMS. Feedstock costs reflect credits for co-products (livestock feed, corn oil, etc.). Feedstock supplies and costs reflect the competition between corn and its co-products and alternative crops, such as soybeans and their co-products.

Cellulosic Biomass feedstock supplies and costs are taken from the NEMS Renewable Fuels Model. Capital and operating costs for biomass ethanol are derived from an Oak Ridge National Laboratory report.[104]

- Current U.S ethanol production capacity is aggregated by census division in the PMM. A small amount of Carribean imports into Census Division 9 is also assumed Cellulose ethanol demonstration plants are modeled in Census Divisions 2 and 7. However, the majority of cellulose ethanol growth is projected in Census Divisions 3 and 4 using corn stover as feedstock, and in Census Division 9 with rice straw and forest residue as the primary feedstock.

- The tax subsidy to ethanol of $0.54 per gallon of ethanol (5.4 cents per gallon subsidy to gasohol at a 10-percent volumetric blending portion) is applied within the premium. This subsidy is scheduled to be reduced to 51 cents by 2007. The tax subsidy is held constant in nominal terms, decreasing with inflation throughout the forecast. The subsidy is assumed not to expire during the forecast period.

- Interregional transportation is assumed to be by rail, and the associated costs are included in the Petroleum Market Model.

The Tax Payer Relief Act of 1997 reduced excise taxes on liquefied petroleum gases and methanol produced from natural gas. The reductions set taxes on these products equal to the Federal gasoline tax on a Btu basis.

With a goal of reducing tailpipe emissions in areas failing to meet Federal air quality standards (nonattainment areas), Title II of the Clean Air Act Admendments of 1990 (CAAA90) established regulations for gasoline formulation. Starting in November 1992, gasoline sold during the winter in the initial 39 carbon monoxide nonattainment areas was required to be oxygenated.[105] Starting in 1995, gasoline sold in major U.S. cities that are considered the most severe ozone nonattainment areas must be reformulated to reduce volatile organic compounds (which contribute to ozone formation) and toxic air pollutants, as well as meet a number of other new specifications. Additional areas with less severe ozone problems have chosen to "opt in" to the reformulated gasoline requirement. In 1998 reformulated gasoline because required to meet a performance based definition, "The Complex Model". In 2000 the performance measures will become more stringent.

Title II of the CAAA90 also established regulations on the sulfur and aromatics content of diesel fuel, which took effect October 1, 1993. All diesel fuel sold for use on highways now contains less sulfur and meets new aromatics or cetane level standards.

A number of pieces of legislation are aimed at controlling air, water, and waste emissions from refineries themselves. The PMM incorporates related environmental investments as refinery fixed costs. The estimated expenditures are based on results of the 1993 National Petroleum Council Study.[106]These investments reflect compliance with Titles I, III, and V of CAAA90, the Clean Water Act, the Resource Conservation and Recovery Act, and anticipated regulations including the phaseout of hydrofluoric acid and a broad-based requirement for corrective action. No costs for remediation beyond the refinery site are included.

Lifting the ban on exporting Alaskan crude oil was passed and signed into law (PL 104-58) in November 1995. Alaskan exports of crude oil have represented about 60 percent of U.S. crude oil exports since November 1995 and are assumed to equal 60 percent of total U.S. crude oil exports in the forecast.

Reduced Sulfur Gasoline Cases

The regulations for Tier 2 emissions standards and related sulfur reductions for gasoline and diesel fuel have not been finalized and are therefore not included in the *AEO2000* reference case. The potential impacts of these proposed regulations are explored in an alternative *gasoline sulfur reduction case*. (Table 64) This case assumes a reduction in gasoline sulfur content to 30 ppm. The 30 ppm limit is met by all reformulated gasoline by 2004. Conventional gasoline is substantially reduced, to 80 ppm in 2004 and meets the 30 ppm restriction by 2007. The more gradual reduction for conventional gasoline reflects extensions that will be granted to small refiners. In order to reduce gasoline sulfur to the level of 30 ppm, refiners will need to invest in conventional hydrotreating processes or in newly developed desulfurization processes, which are potentially less costly but commercially unproven. *AEO2000* included a national low-sulfur gasoline case that did not include new desulfurization technologies. Unlike the low-sulfur scenario in *AEO2000,* this year's gasoline sulfur reduction case incorporates new desulfurization technologies. In the gasoline sulfur reduction case, gasoline consumption and crude oil price projections remain the same as in the *AEO2000* reference case. For consistency with other recent cost analyses, the sulfur reduction case uses a 15-percent hurdle rate and a 10-percent return on investment, and the results are compared with those of a modified reference case using the same financial assumptions.

Table 64. Specifications for Gasoline Sulfur Reduction Scenario

PADD	Reid Vapor Pressure (Max)	Oxygen Weight Percent (Min)	(Max)	Aromatics Volume Percent (Max)	Benzene Volume Percent (Max)	Sulfur Per PPM (Max)	Olefin Volume Percent (Max)	Percent Evaporated at 200°	Percent Evaluated at 300°
TraditionalGasoline PADD I-IV									
2000	10	—	—	28.6	1.60	338.40	10.8	41	83
2001	10	—	—	28.6	1.60	256.63	10.8	41	83
2002	10	—	—	28.6	1.60	187.44	10.8	41	83
2003	10	—	—	28.6	1.60	130.83	10.8	41	83
2004	10	—	—	28.6	1.60	86.80	10.8	41	83
2005	10	—	—	28.6	1.60	55.35	10.8	41	83
2006	10	—	—	28.6	1.60	36.46	10.8	41	83
2007-2020	10	—	—	28.6	1.60	30.00	10.8	41	83
PADD V									
2000	9.2	—	—	28.6	1.60	338.40	10.8	41	83
2001	9.2	—	—	28.6	1.60	256.63	10.8	41	83
2002	9.2	—	—	28.6	1.60	187.44	10.8	41	83
2003	9.2	—	—	28.6	1.60	130.83	10.8	41	83
2004	9.2	—	—	28.6	1.60	86.50	10.8	41	83
2005	9.2	—	—	28.6	1.60	55.35	10.8	41	83
2006	9.2	—	—	28.6	1.60	36.46	10.8	41	83
2007-2020	9.2	—	—	28.6	1.60	30.00	10.8	41	83
Reformulated Gasoline PADD I-IV									
2000	8.5	2.1	2.1	25	0.95	135.00	12	49	87
2001	8.5	2.1	2.1	25	0.95	108.75	12	49	87
2002	8.5	2.1	2.1	25	0.95	82.50	12	49	87
2003	8.5	2.1	2.1	25	0.95	56.25	12	49	87
2004-2020	8.5	2.1	2.1	34	0.95	30.00	12	49	85
PADD V									
2000-2020	7.9	1.7	1.8	25	0.72	25	6	49	85

Source: Energy Information Administration, Office of Integrated Analysis and Forecasting.

Reduced Methyl Tertiary Butyl Ether Case

This alternative case reflects recommendations from a Blue Ribbon Panel (BRP) of experts convened by the EPA to study problems associated with methyl tertiary butyl ether (MTBE) in water supplies. In addition to tighter controls on leaking underground storage tanks, the BRP recommend a substantial reduction in MTBE in gasoline and removal of the Federal oxygen requirement for reformulated gasoline. The BRP further noted that other ethers, such as ethyl tertiary butyl ether (ETBE) and tertiary amyl methyl ether (TAME), have similar but not identical characteristics and recommended studying the health effects and characteristics of those compounds before they are allowed to be placed in widespread use. Because of the greater scrutiny, refiners and blenders are unlikely to increase the use of these ethers significantly. As a result, the use of all ethers in gasoline was assumed to be limited in this case. Although the BRP did not specify a target level of MTBE, but only stated that its use should be reduced substantially, the level of MTBE and other ethers in gasoline was limited to 3 percent by volume in this case. This was the level of refinery inputs of MTBE in gasoline in 1993, the first year in which EIA published the MTBE inputs separately. The use of MTBE began to increase as a result of the introduction of oxygenated gasoline in the fall of 1993. The elimination of the oxygen specification in RFG requires that other specifications be adjusted in order to maintain air quality. In order to maintain current emissions levels of air toxics, as recommended by the BRP, the *BRP/MTBE reduction case* assumes tighter limits on benzene and sulfur in RFG than does the *AEO2000* reference case (Table 65). In the MTBE reduction case, gasoline consumption and crude oil price projections remain the same as in the *AEO2000* reference case. The only changes relative to the reference case are gasoline specifications and the cap on ether use. For consistency with other recent cost analyses, the MTBE reduction case uses a 15-percent hurdle rate and a 10-percent return on investment, and the results are compared with those of a modified reference case using the same financial assumptions.

Table 65. Gasoline Specifications for MTBE Reduction Scenario

PADD	Reid Vapor Pressure (Max)	Oxygen Weight Percent (Min)	(Max)	Aromatics Volume Percent (Max)	Benzene Volume Percent (Max)	Sulfur PPM (Max)	Olefin Volume Percent (Max)	Percent Evaporated at 200°	Percent Evaluated at 300°
Traditional Gasoline **PADD I-IV**									
2000-2002	10.0	—	—	28.6	1.6	338.4	10.8	41.0	83.0
2003-2020	10.0	—	—	27	1.5	338.4	10.8	41.0	83.0
PADD V									
2000-2020	9.2	—	—	28.6	1.6	338.4	10.8	41.0	83.0
Reformulated Gasoline **PADD I-IV**									
2000-2002	8.5	2	2.1	25	0.95	135	10.8	49.0	87.0
2003-2020	8.5	0.005	2.1	30.5	0.55	80	12	49.0	84.0
PADD V									
2000-2002	7.9	1.7	1.8	25.0	0.72	25.0	6.0	49.0	85.0
2003-2020	7.9	0.005	1.8	25.0	0.72	25.0	6.0	49.0	85.0

Source: Energy Information Administration, Office of Integrated Analysis and Forecasting.

Notes and Sources

[100] Energy Information Administration, EIA Model Documentation: Petroleum Market Model of the National Energy Modeling System, DOE/EIA-M059 (2000), January 2000.

[101] Federal Register, Environmental Protection Agency, 40 CFR Part 80, Regulation of Fuels and Fuel Additives: *Standards for Reformulated and Conventional Gasoline, Rules and Regulations,* p. 7800, (Washington, DC, February 1994).

[102] Environmental cost estimates are based on National Petroleum Council, U.S. Petroleum Refining - Meeting Requirements for Cleaner Fuels and Refineries, Volume I, (Washington, DC, August 1993). Associated operating and maintenance base costs predating 1995 are excluded as they are reflected in the refinery fixed operating cost estimates.

[103] American Petroleum Interstate. *"How Much We Pay for Gasoline": 1996 Annual Review,* Page 4 (Washington, DC, May 1997).

[104] M. Walsh, R. Perlock, D. Becker, A Turhollow, and R. Graham, "*Evolution of the Fuel Ethanol Industry: Feedstock Availability and Price*", Oak Ridge National Laboratory (June 5, 1997).

[105] Oxygenated gasoline must contain an oxygen content of 2.7 percent by weight.

[106] National Petroleum Council, U.S. Petroleum Refining - Meeting Requirements for Cleaner Fuels and Refineries, Volume I, (Washington, DC, August 1993).

Coal Market Module

The NEMS Coal Market Module (CMM) provides forecasts of U.S. coal production, consumption, exports, distribution, and prices. The CMM comprises three functional areas: coal production, coal distribution, and coal exports. A detailed description of the CMM is provided in the EIA publication, *Coal Market Module of the National Energy Modeling System 2000*, DOE/EIA-M060(2000) January 2000.

Key Assumptions

Coal Production

The coal production submodule of the CMM generates a different set of supply curves for the CMM for each year of the forecast. Separate supply curves are developed for each of 11 supply regions, and 12 coal types (unique combinations of thermal grade, sulfur content, and mine type). The modeling approach used to construct regional coal supply curves addresses the relationship between the minemouth price of coal and corresponding levels of coal production, labor productivity, and the cost of factor inputs (mining equipment, mine labor, and fuel requirements).

The key assumptions underlying the coal production modeling are:

- Mining costs are assumed to vary with changes in mine production, labor productivity, and factor input costs. Factor input costs are represented by projections of electricity prices from the Electricity Market Module (EMM) and estimates of future coal mine labor and mining equipment costs.

- Between 1978 and 1998, U.S. coal mining productivity (measured in short tons of coal produced per miner per hour) increased at an average rate of 6.7 percent per year. The major factors underlying these gains were interfuel price competition, structural change in the industry, and technological improvements in coal mining.[107] Based on the expectation that further penetration of certain more productive mining technologies, such as longwall methods and large capacity surface mining equipment, will gradually level off, productivity improvements are assumed to continue, but to decline in magnitude. Different rates of improvement are assumed by region and by mine type, surface and underground. On a national basis, labor productivity increases on average at a rate of 2.3 percent a year over the entire forecast, declining from an annual rate of 6.0 percent in 1998 to approximately 1.5 percent over the 2010 to 2020 period. These estimates are based on recent historical data reported on Form EIA-7A, *Coal Production Report*, and expectations regarding the penetration and impact of new coal mining technologies.[108]

- Between 1985 and 1998, the average hourly wage for U.S. coal miners (in 1998 dollars) declined at an average rate of 1.0 percent per year, falling from $21.87 to $19.15.[109] In the reference case, the wage rate for U.S. coal miners, is assumed to remain constant in 1998 dollars (i.e., increase at the general rate of inflation), as it has since 1993.

Coal Distribution

The coal distribution submodule of the CMM determines the least-cost (minemouth price plus transportation cost) supplies of coal by supply region for a given set of coal demands in each demand sector in each demand region using a linear programming algorithm. Production and distribution are computed for 11 supply and 13 demand regions for 18 demand subsectors.

The projected levels of industrial, coking, and residential/commercial coal demand are provided by the industrial, commercial, and residential demand modules; electricity coal demands are provided by the EMM, and coal export demands are provided from the CMM itself.

The key assumptions underlying the coal distribution modeling are:

- Base-year transportation costs are estimates of average transportation costs for each origin-destination pair. These costs are computed as the difference between the average delivered price for a demand region (by sector and for export) and the average minemouth price for a supply curve. Delivered price data are from Form EIA-3, *Quarterly Coal Consumption Report-Manufacturing Plants*, Form EIA-5, *Coke Plant Report-Quarterly*, Federal Energy Regulatory Commission (FERC) Form 423, *Monthly Report of Cost and Quality of Fuels for Electric Plants*, and the U.S. Bureau of the Census' Monthly Report EM-545. Minemouth price data are from Form EIA-7A, *Coal Production Report*.

- Coal transportation costs are modified over time in response to projected variations in reference case fuel costs (No. 2 diesel fuel in the industrial sector), labor costs, the producer price index for transportation equipment, and a time trend. The transportation rate multipliers used for all five *AEO2000* cases are shown in Table 66.

- Electric utility demand received by the CMM is subdivided into "coal groups" representing demands for different sulfur and thermal heat content categories. This process allows the CMM to determine the economically optimal blend of different coals to minimize delivered cost, while meeting the sulfur emissions requirements of the Clean Air Act Amendments of 1990. Similarly, nonutility demands are subdivided into subsectors with their own coal groups to ensure that, for example, lignite is not used to meet a coking coal demand.

Table 66. Transportation Rate Multipliers
(1998=1.000)

Year	Reference Case	High Oil Price	Low Oil Price	High Economic Growth	Low Economic Growth
1998	1.0000	1.0000	1.0000	1.0000	1.0000
2000	1.0340	1.0441	1.0245	1.0437	1.0531
2005	1.0199	1.0291	1.0023	1.0216	1.0158
2010	0.9671	0.9791	0.9503	0.9836	0.9593
2015	0.8962	0.9104	0.8791	0.9223	0.8828
2020	0.8216	0.8320	0.8042	0.8478	0.7991

Source: Energy Information Administration, Office of Integrated Analysis and Forecasting.

Coal Exports

Coal exports are modeled as part of the CMM's linear program that provides annual forecasts of U.S. steam and metallurgical coal exports, in the context of world coal trade. The linear program determines the pattern of world coal trade flows that minimize the production and transportation costs of meeting a prespecified set of regional world coal import demands. It does this subject to constraints on export capacity and trade flows.

The CMM projects steam and metallurgical coal trade flows from 16 coal-exporting regions of the world to 20 import regions for three coal types (coking, bituminous steam, and subbituminous). It includes five U.S. export regions and four U.S. import regions.

The key assumptions underlying coal export modeling are:

- The coal market is competitive. In other words, no large suppliers or groups of producers are able to influence the price through adjusting their output. Producers' decisions on how much and who they supply are driven by their costs, rather than prices being set by perceptions of what the market can bear. In this situation, the buyer gains the full consumer surplus.

- Coal buyers (importing regions) tend to spread their purchases among several suppliers in order to reduce the impact of potential supply disruption, even though this adds to their purchase costs. Similarly, producers choose not to rely on any one buyer and instead endeavor to diversify their sales.

- Coking coal is treated as homogeneous. The model does not address quality parameters that define coking coals. The values of these quality parameters are defined within small ranges and affect world coking coal flows very little.

Data inputs for coal export modeling:

- U.S. coal exports are determined, in part, by the projected level of world coal import demand. World steam and metallurgical coal import demands for the *AEO2000* forecast cases are shown in Tables 67 and 68.

- Step-function coal export supply curves for all non-U.S. supply regions. The curves provide estimates of export prices per metric ton, inclusive of minemouth and inland freight costs, as well as the capacities for each of the supply steps.

- Ocean transportation rates (in dollars per metric ton) for feasible coal shipments between international supply regions and international demand regions. The rates take into account maximum vessel sizes that can be handled at export and import piers and through canals and reflect route distances in thousand nautical miles.

Legislation

It is assumed that provisions of the Energy Policy Act of 1992 that relate to the future funding of the Health and Benefits Fund of the United Mine Workers of America will have no significant effect on estimated production costs, although liabilities of company's contributions will be redistributed. Electricity sector demand for coal, which represented 90 percent of domestic coal demand in 1998, incorporates the provisions of the Clean Air Act Amendments of 1990. It is assumed that electricity producers will be granted full flexibility to meet the specified reductions in sulfur dioxide emissions.

Climate Change Action Plan

Provisions of the Climate Change Action Plan (CCAP) that concern coalbed methane recovery are incorporated in the Oil and Gas Supply Module.

Mining Cost Cases

In the reference case, labor productivity is assumed to increase at an average rate of 2.3 percent a year through 2020, while wage rates remain constant in 1998 dollars. Two alternative cases were modeled in the NEMS CMM, assuming different growth rates for both labor productivity and miner wages. In a low mining cost sensitivity case, productivity increases at 3.6 percent a year, and real wages decline by 0.5 percent a year. In a high mining cost sensitivity case, productivity increases by only 0.9 percent a year, and real wages increase by 0.5 percent a year. In the alternative cases, the annual growth rates for productivity were increased and decreased by mine type (underground and surface), based on historical variations in labor productivity during the years 1980 through 1997. Both cases were run using only the NEMS Energy Supply Modules (Oil and Gas Supply Module, Natural Gas Transmission and Distribution Module, Coal Market Module, and Renewable Fuels Module), the Petroleum Market Module, and the Electricity Market Module, rather than as a fully integrated NEMS run. Consequently, no price-induced demand feedback in end-use coal markets was captured. In an integrated run, the demand response would tend to moderate the magnitude of the equilibrium price response.

Table 67. World Steam Coal Import Demand by Import Region, 1998-2020
(Million Metric Tons of Coal Equivalent)

Import Regions[1]	1998	2005	2010	2015	2020
The Americas	29.0	32.7	33.6	35.1	36.6
United States	7.4	13.3	14.8	15.9	16.9
Canada	12.7	8.6	7.4	7.6	8.2
Mexico	3.1	6.0	6.5	6.7	6.7
South America	5.8	4.8	4.9	4.9	4.9
Europe	106.2	111.0	108.4	104.3	101.1
Scandinavia	12.6	8.5	5.6	4.4	3.6
U.K./Ireland	13.9	15.8	16.9	16.9	16.9
Germany	13.5	21.8	21.8	22.7	24.5
Other NW Europe	24.7	20.5	17.7	14.5	10.9
Iberia	14.0	11.0	11.3	11.3	9.5
Italy	8.7	8.7	8.3	7.8	7.3
Med/E Europe	18.8	24.7	26.8	26.7	28.4
Asia	149.9	200.5	232.0	245.7	261.5
Japan	63.8	80.2	90.6	92.8	94.2
East Asia	62.2	81.1	91.0	95.5	99.1
China/Hong Kong	8.3	11.7	16.2	20.7	27.0
ASEAN	9.2	21.6	26.8	28.4	31.1
Indian Sub	6.4	5.9	7.4	8.3	10.1
Total	285.1	344.2	374.0	385.1	399.2

[1]Import Regions: **United States:** United States; **Canada:** Canada; **Mexico:** Mexico; **South America:** Argentina, Brazil, Chile; **Scandinavia:** Denmark, Finland, Norway, Sweden; **U.K./Ireland:** Ireland, United Kingdom; **Germany:** Austria, Germany; **Other NW Europe:** Belgium, France, Luxembourg, Netherlands; **Iberia:** Portugal, Spain; **Italy:** Italy; **Med/E Europe:** Algeria, Bulgaria, Croatia, Egypt, Greece, Israel, Malta, Morocco, Romania, Tunisia, Turkey; **Japan:** Japan; **East Asia:** North Korea, South Korea, Taiwan; **China/Hong Kong:** China, Hong Kong; **ASEAN:** Malaysia, Philippines, Thailand; **Indian Sub:** Bangladesh, India, Iran, Pakistan, Sri Lanka.

Source: Energy Information Administration, Office of Integrated Analysis and Forecasting.

Notes: One "metric ton of coal equivalent" contains 27.78 million Btu. Totals may not equal sum of components due to independent rounding.

Table 68. World Metallurgical Coal Import Demand by Import Region, 1998-2020
(Million Metric Tons of Coal Equivalent)

Import Regions[1]	1998	2005	2010	2015	2020
The Americas	19.9	23.1	25.2	28.3	30.2
United States	0.5	0.5	0.5	0.5	0.5
Canada	4.7	4.4	4.2	4.0	3.7
Mexico	0.3	2.2	2.8	3.6	3.8
South America	14.4	16.0	17.7	20.2	22.2
Europe	57.5	54.9	54.4	53.7	52.9
Scandinavia	3.1	2.7	2.4	2.2	1.9
U.K./Ireland	9.2	7.6	7.6	7.1	7.1
Germany	4.2	6.9	6.9	6.9	6.9
Other NW Europe	17.3	15.1	13.2	12.2	11.2
Iberia	4.7	3.8	3.8	3.8	3.8
Italy	7.6	7.2	7.1	6.3	6.3
Med/E Europe	11.4	11.6	13.4	15.2	15.7
Asia	95.9	99.9	102.2	104.8	106.9
Japan	60.9	56.4	52.7	51.3	49.9
East Asia	23.5	28.4	31.7	33.6	35.9
China/Hong Kong	0.1	0.6	1.7	1.7	1.9
ASEAN	0.0	0.0	0.0	0.0	0.0
Indian Sub	11.4	14.5	16.1	18.2	19.2
Total	173.3	177.9	181.8	186.8	190.0

[1]Import Regions: **United States:** United States; **Canada:** Canada; **Mexico:**; **South America:** Argentina, Brazil, Chile; **Scandinavia:** Denmark, Finland, Norway, Sweden; **U.K./Ireland:** Ireland, United Kingdom; **Germany:** Austria, Germany; **Other NW Europe:** Belgium, France, Luxembourg, Netherlands; **Iberia:** Portugal, Spain; **Italy:** Italy; **Med/E Europe:** Algeria, Bulgaria, Croatia, Egypt, Greece, Israel, Malta, Morocco, Romania, Tunisia, Turkey; **Japan:** Japan; **East Asia:** North Korea, South Korea, Taiwan; **China/Hong Kong:** China, Hong Kong; **ASEAN:** Malaysia, Philippines, Thailand; **Indian Sub:** Bangladesh, India, Iran, Pakistan, Sri Lanka.

Source: Energy Information Administration, Office of Integrated Analysis and Forecasting.

Notes: One "metric ton of coal equivalent" contains 27.78 million Btu. Totals may not equal sum of components due to independent rounding.

Coal Quality

Each year the values of base year coal production, heat content, sulfur and carbon emissions for each coal source in the Coal Market Module of the NEMS are calibrated to survey data. Surveys used for this purpose are the FERC Form 423, a survey of the origin, cost and quality of fossil fuels delivered to electric utilities, the Form FERC 867 which records the quality of coal receipts at independent power producers, the Form EIA5 and 5a which record the origin, cost, and quality of coal receipts at domestic coke plants, and the Forms EIA 3 and 3a, which record the origin, cost and quality of coal delivered to domestic industrial consumers. Estimates of coal quality for the export and residential/commercial sectors are made using the survey data for coal delivered to coking coal and industrial steam coal consumers. Sulfur emissions levels shown below (Table 69) have been adjusted from percent by weight data (on an 'as received basis') to reflect U.S. Environmental Protection Agency (U.S.E.P.A.) assumptions about the variation in the completeness of combustion by coal rank: 95 percent for bituminous coals, 87.5 percent for sub-bituminous coals and 75 percent for lignite. These emissions are appropriate for unscrubbed boilers; depending on the type of scrubber employed, emissions from scrubbed boilers would be further reduced by between 70 and 95 percent. Carbon emissions levels for each coal type are listed below in pounds of carbon dioxide emitted per million Btu.

Table 69. Production, Heat Content, and Sulfur and Carbon Emissions by Coal Type and Region

Coal Supply Region	Coal Rank & Sulfur Level	1998 Production (Million tons)	Million Btu per Short ton	Lbs SOX Emitted per million Btu	Lbs CO2 Emitted per million Btu
North Appalachia	Metallurgical[1]	5.386	26.80	1.39	205.4
	Low Sulfur Bituminous	2.821	25.48	0.99	203.6
	Mid Sulfur Bituminous	80.842	25.51	2.39	205.4
	High Sulfur Bituminous	68.701	24.45	4.88	203.6
	Waste Coal (Bituminous)	9.500	13.61	3.65	203.6
Central Appalachia	Metallurgical[1]	59.746	26.80	1.03	203.8
	Low Sulfur Bituminous	72.736	25.40	1.12	203.8
	Mid Sulfur Bituminous	144.448	25.10	1.63	203.8
Southern Appalachia	Metallurgical[1]	5.823	26.80	1.03	203.3
	Low Sulfur Bituminous	7.526	24.93	0.95	203.3
	Mid Sulfur Bituminous	12.358	24.60	2.19	203.3
East Interior	Mid Sulfur Bituminous	36.236	23.04	2.30	201.4
	High Sulfur Bituminous	73.940	22.56	5.11	201.4
West Interior/Gulf	High Sulfur Bituminous	2.398	23.37	4.85	202.4
	Mid Sulfur Lignite	28.943	12.94	2.03	211.4
	High Sulfur Lignite	26.856	12.79	3.21	211.4
Dakota Lignite	Mid Sulfur Lignite	30.860	13.30	3.83	216.6
Powder R., Green R. & Hannah Basins	Low Sulfur Subbituminous	318.840	17.37	0.68	210.7
	Mid Sulfur Subbituminous	36.357	17.69	1.33	210.7
	Low Sulfur Bituminous	1.723	21.60	1.00	204.4
Rocky Mountain	Low Sulfur Bituminous	45.780	23.14	0.80	203.0
	Low Sulfur Subbituminous	9.926	20.89	.74	210.6
Southwest	Low Sulfur Bituminous	26.755	21.38	.87	205.4
	Mid Sulfur Subbituminous	13.157	18.42	1.45	206.7
Northwest	Mid Sulfur Subbituminous	5.983	16.34	1.38	207.9

[1]The average heat content of metallurgical coal is estimated to be 26.80 million Btu per short ton in EIA, Table 5, *Annual Energy Review*, DOE/EIA-0384(98), (Washington, DC, July 1999).

Note: Coal Supply Regions are defined as follows: North Appalachia - PA, MD, OH, No. WV; Central Appalachia - So. WV, VA, East KY; South Appalachia - AL, TN; East Interior - IL., IN, West KY; West Interior/Gulf - IA, MO, KS, OK, AR, LA, TX; Dakota lignite - ND, MT (lignite only); Powder River, Green River and Hannah River Basins - MT, WY (subbituminous and bituminous); Rocky Mountain - CO, UT; Southwest-AZ, NM; Northwest - AK, WA.

Source: Energy Information Administration, Office of Integrated Analysis and Forecasting.

Notes and Sources

[107] Energy Information Administration, *The U.S. Coal Industry, 1970-1990: Two Decades of Change*, DOE/EIA-0559, (Washington, DC, November 1992).

[108] Stanley C. Suboleski, et.al., *Central Appalachia: Coal Mine Productivity and Expansion*, Electric Power Research Institute, EPRI IE-7117, (September 1991).

[109] U.S. Department of Labor, Bureau of Labor Statistics.

Renewable Fuels Module

The NEMS Renewable Fuels Module (RFM) consists of five distinct submodules that represent the major renewable energy technologies. Although it is described here, conventional hydroelectric is included in the Electricity Market Module (EMM) and is not part of the RFM. Similarly, ethanol modeling is included in the Petroleum Market Module (PMM). Some renewables, such as municipal solid waste (MSW) and other biomass materials, are fuels in the conventional sense of the word, while others, such as wind and solar radiation, are energy sources that do not require the production or consumption of a fuel. Renewable technologies cover the gamut of commercial market penetration, from hydroelectric power, which was an original source of electricity generation, to newer power systems using wind, solar, and geothermal energy. In some cases, they require technological innovation to become cost effective or have inherent characteristics, such as intermittency, which make their penetration into the electricity grid dependent upon new methods for integration within utility system plans or upon low-cost energy storage.

The submodules of the RFM interact only with modules outside of the RFM and not with other RFM submodules. These interactions occur through common elements of the model with the Electricity Market Module (EMM). Because of the high level of integration with the EMM, the final outputs (levels of consumption and market penetration over time) for renewable energy technologies are largely dependent upon the EMM.

For *AEO2000*, two increasing costs are superimposed onto the capital costs of renewable energy technologies to represent two phenomena:

- Short-term cost adjustment factors, which increase technology capital costs as a result of rapid U.S. buildup in a single year and reflecting limitations on the infrastructure (for example, manufacturing, resource assessment, construction expertise) to accommodate unexpected demand growth. These short-term factors are invoked when demand for new capacity in any year exceeds 30 percent of the prior year's total U.S. capacity. For every 1 percent increase in total U.S. capacity over the previous year greater than 30 percent, capital costs rise 0.5 percent. These factors apply to biomass, geothermal, municipal solid waste, solar, and wind technologies.

- For biomass and wind only, increased costs are assumed to result from a large cumulative increase in use of one of these resources, reflecting any or all of three general factors: (1) resource degradation, (2) transmission network upgrades, and (3) market factors. Presumably best land resources are used first. Increasing resource use necessitates resort to less efficient land - less accessible, less productive, more difficult to use (e.g, land roughness, slope, terrain variability, or productivity, wind turbulence or wind variability). Second, as capacity increases, especially for intermittent technologies like wind power, existing local and long-distance transmission networks require upgrading, increasing overall costs. Third, market pressures from competing land uses increase costs as cumulative capacity increases, including competition from agricultural or other production alternatives, residential or recreational use, aesthetics, or from broader environmental preferences. As a result, for *AEO2000*, each EMM region's biomass and wind resource estimates are parceled into five cost levels. For biomass, the percentage cost increases that are applied to initial capital costs are 0, 15, 50, 75 and 100 percent for successive increments of the resource. For wind, the increases are 0, 20, 50, 100 and 200 percent respectively. The size of the resource increments vary by technology and region.

For an in-depth discussion of the learning functions, see the EMM section and the background section of the model summary for the Geothermal Electric Submodule. A detailed description of the RFM is provided in the EIA publication, *Renewable Fuels Module of the National Energy Modeling System, Model Documentation 2000*, DOE/EIA-M069(2000), January 2000.

Key Assumptions

Nonelectric Renewable Energy Uses

In addition to projections for renewable energy used in electricity generation, the *AEO2000* contains projections of nonelectric renewable energy uses for industrial and residential wood consumption, solar residential and commercial hot water heating, and residential and commercial geothermal (ground-source) heat pumps. Additional renewable energy applications, such as direct solar thermal industrial applications or direct lighting, off-grid electricity generation, and heat from geothermal resources used directly (e.g., district heating and greenhouses) are not included in the projections.

Electric Power Generation

The RFM specifically and EMM in general consider only grid-connected central station electricity generation. Off-grid and distributed sources, such as off-grid and distributed grid-connected applications of photovoltaic, dish-Stirling solar, and wind generation, are not included in the energy projections for the *AEO2000*. The renewable submodules that interact with the EMM are the central station grid-connected solar (thermal and photovoltaic), wind, geothermal, biomass, and MSW submodules. Most provide specific data that characterize that resource in a useful manner. In addition, a set of technology cost and performance values is provided directly to the EMM. These data are central to the build and dispatch decisions of the EMM with the exception of MSW. The values are presented in Table 37 of the EMM section.

Conventional Hydroelectricity

Background

The Hydroelectric Power Data File in the EMM represents reported plans for new conventional hydroelectric power capacity connected to the transmission grid and reported on Form EIA-860, *Annual Electric Generator Report*, and Form EIA-867, *Annual Nonutility Power Producer Report*. It does not estimate pumped storage hydroelectric capacity, which is considered a storage medium for coal and nuclear power and not a renewable energy use. However, the EMM allows new conventional hydroelectric capacity to be built in addition to reported plans. Converting Idaho National Engineering and Environmental Laboratory information on U.S. Hydroelectric potential, the EMM contains regional conventional hydroelectric supply estimates at increasing capital costs. All the capacity is assumed available at a uniform capacity factor of 45 percent. Data maintained for hydropower include the available capacity, capacity factors, and costs (capital, and fixed and variable operating and maintenance). The fossil-fuel heat rate equivalents for hydropower are provided to the report writer for energy consumption calculation purposes only.

Assumption

- Because of hydroelectric power's position in the merit order of generation, it is assumed that all available installed hydroelectric capacity will be used within the constraints of available water supply and general operating requirements (including environmental regulations).

Solar Electric Submodule

Background

The Solar Electric Submodule (SOLES) currently includes two solar technologies: 50 megawatt central receiver (power tower) solar thermal (ST) and 5 megawatt fixed-flat plate thin-film copper-indium-diselenide (CIS) photovoltaic (PV) technologies. PV is assumed available in all thirteen EMM regions, while ST is available only in the six primarily Western regions where direct normal solar insolation is sufficient. Capital costs for both technologies are determined by EIA using multiple sources, including technology characterizations by the Department of Energy's Office of Energy Efficiency and Renewable Energy and the Electric Power Research Institute (EPRI).[110] Most other cost and performance characteristics for ST are obtained or derived from the August 6, 1993, California Energy Commission memorandum, *Technology Characterization for ER 94*; and, for PV, from the Electric Power Research Institute, *Technical Assessment*

Guide (TAG) 1993.   In addition, capacity factors are obtained from information provided by the National Renewable Energy Laboratory (NREL).

Assumptions

- Capacity factors for solar technologies are assumed to vary by time of day and season of year, such that nine separate capacity factors are provided for each modeled region, three for time of day and for each of three broad seasonal groups (summer, winter, and spring/fall).   The current solar thermal annual capacity factor for the region including California, for example, is assumed to average 40 percent; California's current PVcapacity factor is assumed to average 24.6 percent.

- In order to incorporate assumed improvements in photovoltaic technologies, all PV capacity factors are assumed to improve linearly a total of 10 percent from 2005 through 2015; for example, California's annual average capacity factor for PV increases from 24.6 percent to almost 27.1 percent by 2015.

- Because solar technologies are more expensive than other utility grid-connected technologies, early penetration will be driven by broader economic decisions such as the desire to become familiar with a new technology or environmental considerations.   Early year's penetration for such reasons is included as supplemental additions.

- Solar resources are well in excess of conceivable demand for new capacity; therefore, energy supplies are considered unlimited within regions (at specified daily, seasonal, and regional capacity factors).   Accordingly, there is no reason to track solar resources in NEMS.   In the seven regions where ST technology is not modeled, the level of direct, normal insolation (the kind needed for that technology) is insufficient to make that technology commercially viable through 2020.

- NEMS models the Energy Policy Act of 1992 (EPACT) 10-percent investment tax credit for solar electric power generation by tax-paying entities.

Wind-Electric Power Submodule

Background

Because of limits to windy land area, wind is considered a finite resource, so the submodule calculates maximum available capacity by Electricity Market Module Supply Regions.   The minimum economically viable wind speed is about 13 mph, and wind speeds are categorized into three wind classes according to annual average wind speed. The RFM keeps track of wind capacity (megawatts) within a region and moves to the next best wind class when one category is exhausted. Wind resource data on the amount and quality of wind per EMM region come from a Pacific Northwest Laboratory study and a subsequent update.[111]   The technological performance, cost, and other wind data used in NEMS are derived by EIA from consultation with industry experts.[112] Maximum wind capacity, capacity factors, and incentives are provided to the EMM for capacity planning and dispatch decisions. These form the basis on which the EMM decides how much power generation capacity is available from wind energy. The fossil-fuel heat rate equivalents for wind are provided to the report writer for energy consumption calculation purposes only.

Assumptions

- Only grid-connected (utility and nonutility) generation is included.   The forecasts do not include off-grid or distributed electric generation.

- In the wind submodule, wind supply is constrained by three modeling measures, addressing (1) average wind speed, (2) distance from existing transmission lines, and (3) resource degradation, transmission network upgrade costs, and market factors.

- First availability of wind power (among three wind classes) is based on the Pacific Northwest Laboratory Environmental and Moderate Land-Use Exclusions Scenario, in which some of the windy land area is not available for siting of wind turbines. The percent of total windy land unavailable under this scenario consists of all environmentally protected lands (such as parks and wilderness areas), all urban lands, all wetlands, 50 percent of forest lands, 30 percent of agricultural lands, and 10 percent of range and barren lands.

- Wind resources are mapped by distance from existing transmission capacity among three distance categories, accepting wind resources within (1) 0-5, (2) 5-10, and (3) 10-20 miles on either side of the transmission lines. Transmission cost factors are added to the resources further from the transmission lines.

- For *AEO2000*, capital costs for wind technologies are also assumed to increase in response to (1) declining natural resource quality, such as terrain slope, terrain roughness, terrain accessibility, wind turbulence, wind variability, or other natural resource factors, (2) increasing cost of upgrading existing local and network distribution and transmission lines to accommodate growing quantities of intermittent wind power, and (3) market conditions, the increasing costs of alternative land uses, including for aesthetic or environmental reasons. Capital costs are left unchanged for some initial share, then increased 20, 50, 100 percent, and finally 200 percent, to represent the aggregation of these factors. Proportions in each category vary by EMM region.

- Depending on the EMM region, the cost of competing fuels and other factors, wind plants can be built to meet system capacity requirements or as "fuel savers" to displace generation from existing capacity. For wind to penetrate as a fuel saver, its total capital and fixed operations and maintenance costs minus applicable subsidies from the EPACT, must be less than the variable operating and fuel costs for existing (non-wind) capacity.

- Because of downwind turbulence and other aerodynamic effects, the model assumes an average spacing between turbine rows of 5 rotor diameters and a lateral spacing between turbines of 10 rotor diameters. This spacing requirement determines the amount of power that can be generated from windy land area and is factored into requests for generating capacity by the EMM.

- It is expected that wind turbine technology will improve in performance and that blade lengths will increase, as the cubic relationship between the area swept by the rotor and power generation provides a large incentive for increasing blade length. Capacity factors are assumed to increase to a national average of about 34 percent in the best wind class. However, as better wind resources are depleted, capacity factors are assumed to go down.

Geothermal-Electric Power Submodule

Background

In developing geothermal capacity growth projections, the focus is on hydrothermal resources; extraction of energy from hot dry rock resources is not included. This is because the technology probably will at best be available after 2010, and reliable cost and resource data are not available. The Geothermal-Electric Power Submodule (GES) utilizes a process of resource accounting based on Sandia National Laboratory's 1991 geothermal resource assessment.[113] Site-specific costs, including those for drilling, steam collection, and electricity transmission to the grid, as well as site characteristics, are used in identifying available resources and capacities by EMM region. The cost and performance values are based on dual flash and binary cycle technologies. The costs from 51 sites are aggregated into a set of regional supply curves for each year. For each iteration of a model run, a value for avoided cost is obtained from the Electricity Capacity Planning Submodule to establish the levelized cost at which to truncate the curves, thereby excluding the higher cost resources. Capital cost learning on the generating units which emulates what is done in the EMM is incorporated in the GES.[114] For *AEO2000*, the capacity factor has been set at 87 percent, based on historical data.

Assumptions

- Existing and planned capacity data are accessed directly by the EMM. The data are obtained from Forms EIA-860 and EIA-867.

- An investment tax credit of 10 percent is assumed to be available in all forecast years based on the EPACT.

- Plant retirements are generally assumed to occur 30 years after startup. An exception is made for wells affected by a project to bring water to parts of The Geysers site which is expected to halt the enthalpy decline occurring there. Of these (six) wells, half are assumed to be retired after 35 years, the others in 40 years.

- Capital and operating costs vary by site and year; values shown in Table 37 are indicative of those used by EMM for geothermal build and dispatch decisions.

Biomass Electric Power Submodule

Background

Biomass consumed for electricity generation is modeled in two parts in NEMS. Capacity in the wood products and paper industries, the so-called captive capacity, is included in the industrial sector module as cogeneration. Generation by the electricity sector is represented in the EMM, with capital and operating costs and capacity factors as shown in Table 37, as well as fuel costs, being passed to the EMM where it competes with other sources. Fuel costs are provided in sets of regional supply schedules. Projections for ethanol are produced by the Petroleum Market Module (PMM), with the quantities of biomass consumed for ethanol decremented from, and prices obtained from, these same supply schedules.

Assumptions

- Existing and planned capacity data are accessed directly by the EMM. The data are obtained from Forms EIA-860 and EIA-867.

- The conversion technology represented, upon which the costs in Table 37 are based, is an advanced gasification-combined cycle plant that is similar to a coal-fired gasifier. Costs in the reference case were developed by EIA to be consistent with coal gasifier costs.

- Biomass cofiring can occur up to a maximum of 5 percent of fuel used in coal-fired generating plants. However, co-firing is not enumerated as biomass capacity but remains accounted as part of coal-fired capacity.

In place of the previously used capacity constraints, short-term and long-term cost adjustment factors have been installed. These values are described in the RFM introduction.

Fuel supply schedules are a composite of four fuel types; forestry materials, wood residues, agricultural residues and energy crops. The first three are combined into a single supply schedule for each region which does not change for the full forecast period. Energy crops data are presented in yearly schedules from 2010 to 2020 in combination with the other material types for each region. The forestry materials component is made up of logging residues, rough rotten salvable dead wood and excess small pole trees.[115] The wood residue component consists of primary mill residues, silvicultural trimmings and urban wood such as pallets, construction waste and demolition debris that are not otherwise used.[116] Agricultural residues are wheat straw and corn stover only, which make up the great majority of crop residues.[117] Energy crops data is for hybrid poplar, willow and switchgrass grown on crop land, pasture land, or on Conservation Reserve lands.

Energy Crop costs range from zero to over five dollars per million Btu.[118] The maximum amount of resources in each supply category is shown in Table 70.

Municipal Solid Waste-Electric Power Submodule

Table 70. U.S Biomass Resources, by Region and Type, 2020
(Trillion Btu)

	Forest Resources	Mill Residue	Energy Crops	Agricultural Residue	Total
1. ECAR	780	69	850	422	2121
2. ERCOT	126	28	264	59	478
3. MAAC	267	33	103	37	440
4. MAIN	204	22	819	454	1499
5. MAPP	260	16	2306	982	3564
6. NPCC/NY	272	26	92	11	401
7. NPCC/NE	466	40	26	0	532
8. SERC/FL	124	24	14	5	167
9. SERC	1608	94	669	67	2438
10. SPP	795	125	1365	274	2559
11. NWP	1097	41	32	55	1224
12. W/RA	195	20	0	55	270
13. W/CNV	346	59	0	24	429
Total US	**6541**	**597**	**6541**	**2444**	**16122**

Source: Energy Information Administration, Office of Integrated Analysis and Forecasting.

Background

Municipal solid waste (MSW) combustion is treated within NEMS as a separate technology whose electricity production is determined exogenous to the EMM. The cost of producing electricity is passed to the EMM only as an input to the calculation that derives the average cost of producing electricity. Energy from MSW is a byproduct of waste disposal activity and, therefore, does not compete against other technologies in model decisions regarding new capacity additions.[119]

Assumptions

- MSW is assumed to displace other energy forms lower in the merit order.

- Build decisions are based on a stepwise process involving waste disposal parameters.

  - Gross domestic product (GDP) and population are used as the drivers in an econometric equation that establishes the supply of MSW.

  - The values disposal parameters are extrapolated from historical Environmental Protection Agency (EPA) values for MSW and multiplied by 1.47 to reflect a broader definition of materials known to be combusted. The factor 1.47 is derived from information in the Biocycle State Survey.[120]

  - The heat content of the MSW is set at 5,190 Btu per pound, and is assumed to remain at that level throughout the projection.

  - The percentage of waste combusted is assumed to remain constant at 11 percent of a growing waste stream. Using the larger Biocycle-based value for generation of the MSW waste stream, the EIA estimated percentage currently combusted equals 12 percent of the alternative EPA estimate of waste.[121]

  - The total energy from MSW projected for the United States is limited to the portion currently used for electricity generation (about 92 percent) and is disaggregated into regions. This regional breakdown is performed by maintaining the projected 1996 distribution of these factors as represented in the EIA database of MSW plants. Steam from MSW is represented in industrial cogeneration.

- An estimate is made of energy produced from landfill gas. Data values are entered in a spreadsheet that considers existing and additional landfills and a profile of gas generation from the waste. It is assumed that the percent of the gas emitted that will be captured for energy conversion will increase from 13 percent in 1995 to 40 percent in 2020, based on an EPA estimate of a tripling of landfills that will capture the gas. The resulting generating capacity is added to the capacity for MSW combustion, is disaggregated by region and passed to the EMM.

Legislation

Energy Policy Act of 1992 (EPACT)

The RFM includes the investment tax and energy production credits called for in the EPACT for the appropriate energy types. EPACT provides a renewable electricity production credit of 1.5 cents per kilowatt-hour for electricity produced by wind, applied to plants that become operational between January 1, 1994, and December 31, 1999; *AEO2000* does not assume that the EPACT credit will be extended. The credit extends for 10 years after the date of initial operation. EPACT also includes provisions that allow an investment tax credit of 10 percent for solar and geothermal technologies that generate electric power. This credit is represented as a 10-percent reduction in the capital costs in the RFM.

Supplemental Capacity Additions

In addition to the reported generating capacity plans from the EIA-860 and EIA-867 and capacity projected through the use of the EMM/RFM, the *AEO2000* also includes 2,897 megawatts additional generating capacity powered by renewable resources. Summarized in Table 71 and detailed in Table 72, some of the capacity represents mandated new capacity required by state laws, EIA estimates for expected new capacity under recent state-enacted renewable portfolio standards (RPS), estimates of winning bids in California's renewables funding program (Assembly Bill 1890), expected new capacity under known voluntary programs, such as "green marketing" efforts, and other reported plans.

Table 71. Post-1999 Supplemental Capacity Additions (Megawatts, Net Summer Capability)

Rationale	Geothermal	Biomass	Municipal[3] Solid Waste	Solar Thermal	Solar Photovoltaic	Wind	Total
Mandates	0.0	118.8	0.0	0.0	0.0	159.0	277.8
RenewablePortfolio Standards	12.1	851.4	1137.5	34.8	10.8	2605.4	4652.0
California AB1890[1]	150.6	7.4	46.5	0.0	0.0	33.6	238.1
Other Reported Plans[2]	0.0	4.8	26.4	0.0	0.0	50.0	81.2
Total	162.7	982.4	1210.4	34.8	10.8	2848.0	5249.1

1Partially supported by funding under California Assembly Bill 1890.

[2]Other non mandated plans, including "Green Marketing" efforts and other activites known to EIA.

[3]includes Landfill Gas.

Source: Energy Information Administration, Office of Integrated Analysis and Forecasting.

Table 72. Planned Post-1999 U.S. Generating Capacity Using Renewable Resources[1]

Technology	Plant Name	Program	State[2]	Net Summer Capability (Megawatts)	Year On Line
Biomass (Including Wood)	Agrilectric	AB1890	California	7.4	2002
	Connecticut RPS	RPS	Connecticut	90.8	2000-2014
	Massachusetts RPS	RPS	Massachusetts	110.0	2001-2015
	Northern States Power	Mandate	Minnesota	118.8	2001-2002
	New Jersey RPS	RPS	New Jersey	650.7	2003-2015
	Browning Ferris Concord	Commercial	North Carolina	4.8	2000
Geothermal	Imperial Valley	AB1890	California	9.5	2000
	Salton Sea	AB1890	California	46.6	2000
	Four Mile Hill	AB1890	California	47.4	2001
	Telephone Flats	AB1890	California	45.6	2001
	Nevada RPS	RPS	Nevada	13.6	2001-2014
Municipal Solid Waste	Minnesota Methane, LFG	AB1890	California	3.2	2000
	Wheelabrator	AB1890	California	3.8	2000
	Browning Ferris, LFG	AB1890	California	19.0	2000
	Energy Development, LFG	AB1890	California	11.5	2000
	Sacramento Mun. Utility District, LFG	Commercial	California	7.9	2000
	San Francisco	AB1890	California	1.9	2001
	Riverside, LFG	AB1890	California	7.1	2002
	Denver Allied Waste, LFG	Commercial	Colorado	5.7	2001
	Connecticut RPS	RPS	Connecticut	61.3	2000-2014
	Browning Ferris, Randolph, LFG	Commercial	Massachusetts	2.9	2002
	Browning Ferris, Falls River, LFG	Commercial	Massachusetts	7.0	2002
	Browning Ferris, East Bridgewater, LFG	Commercial	Massachusetts	2.9	2002
	New Jersey RPS, LFG	RPS	New Jersey	650.7	2003-2015
	Texas RPS, LFG	RPS	Texas	425.6	2001-2014
Central Station Solar Photovoltaic	Nevada RPS	RPS	Nevada	10.8	2005-2014
Central Station Solar Thermal	Nevada RPS	RPS	Nevada	34.8	2004-2014
Wind	Mark Technologies	AB1890	California	25.2	2000
	CalWind	AB1890	California	8.4	2000
	Connecticut RPS	RPS	Connecticut	66.0	2000-2014
	Endless Energy Redding	Commercial	Maine	20.0	2001
	Massachusetts RPS	RPS	Massachusetts	366.2	2001-2015

Table 72. Planned Post-1999 U.S. Generating Capacity Using Renewable Resources[1] (Continued)

Technology	Plant Name	Program	State	Net Summer Capacity (Megawatts)	Year On Line
	17 Small Projects	Mandate	Minnesota	25.0	2000
	Northern States Power	Mandate	Minnesota	131.0	2002
	New Jersey RPS	RPS	New Jersey	381.2	2003-2016
	Niagara Mohawk, Lowville	Mandate	New York	3.0	2001
	Green Mountain Power	Commercial	Pennsylvania	10.0	2000
	Austin	Commercial	Texas	20.0	2000
	Texas RPS	RPS	Texas	1792.0	2001-2019

[1]Includes reported information and EIA estimates for mandates, renewable portfolio standards (RPS), and California's renewables.

[2]Support program (AB1890); municipal solid waste estimates identify landfill gas (LFG) where known.

Source: Energy Information Administration, Office of Integrated Analysis and Forecasting.

Climate Change Action Plan

Action Item 26, "*Form Renewable Energy Market Mobilization Collaborative with Technology Demonstration,*" of the Climate Change Action Plan (CCAP),[122] is designed to spur field validation of selected renewable energy technologies by supporting specified electric utility tests. The demonstrations, along with information dissemination, intend to address market barriers by increasing utility and investor confidence in the technologies. Technologies identified in Action Item 26 include assistance to "ice breaker" geothermal plants, site testing advanced wind turbines, and assistance and collaboration in launching test biomass-fueled and photovoltaic electricity generating technologies.

The electricity generating capacity effects on *AEO2000* of Action Item 26 are incorporated in EIA's projections for renewable technologies. The supplemental capacity additions include additions that will be cost-shared by DOE and industry. While the stated goal of this action item is "increased utility and investor experience and confidence" in renewable technologies, in general, no additional cost declines beyond those discussed above are assumed.

Notes and Sources

[110] Electric Power Research Institute and U.S. Department of Energy, Office of Utility Technologies *Renewable Energy Technology Characterizations* (EPRI TR-109496, December 1997) or www.eren.doe.gov/utilities/techchar.html.

[111] D.L. Elliott, L.L. Wendell, and G.L. Gower, *An Assessment of the Available Windy Land Area and Wind Energy Potential in the Contiguous United States*, Richland, WA: Pacific Northwest Laboratory, (August 1991); Pacific Northwest Laboratory, *An Assessment of the Available Windy Land Area and Wind Energy Potential in the Contiguous United States*, (PNL-7789), prepared for the U.S. Department of Energy under Contract DE-AC06-76RLO 1830, (August 1991), and Schwartz, N.N.; Elliot, O.L.; and Gower, GL: *Gridded State Maps of Wind Electric Potential Proceedings Wind Power 1992*, (Seattle, WA, October 19-23, 1992).

[112] Energy Information Administration analysts discussed input values with the Electric Power Research Institute, U.S. Dept. of Energy's Office of Energy Efficiency and Renewable Energy, Lawrence Berkeley National Laboratory, RLA Consulting, and the Zond Corporation.

[113] Sandia National Laboratories, *Supply of Geothermal Power from Hydrothermal Sources: A Study of the Cost of Power in 20 and 40 Years*, (Albuquerque, NM, June 1991).

[114] DynCorp-Meridian Inc., *Model Documentation, Geothermal Electric Submodule of the Renewable Fuels Module of the National Energy Modeling System*, (Alexandria VA, December 1994) and DynCorp Environmental, Energy and National Security Programs Inc., *Model Documentation, Renewable Fuels Module; Modifications to the Geothermal Electricity Supply Submodule*, prepared for the Energy Information Administration, (Alexandria, VA, September 1995).

[115] United States Department of Agriculture, U.S. Forest Service, *Forest Resources of the United States, 1992*, General Technical Report RM-234, (Fort Collins CO, June 1994).

[116] Antares Group Inc., *Biomass Residue Supply Curves for the U.S.*, prepared for the National Renewable Energy Laboratory, November 1998.

[117] Walsh, M.E., et.al., Oak Ridge National Laboratory, *Evolution of the Fuel Ethanol Industry: Feedstock Availability and Price* (Oak Ridge, TN, April 21, 1998).

[118] Graham, R.L., et.al., Oak Ridge National Laboratory, *The Oak Ridge Energy Crop County Level Database*, (Oak Ridge TN, December, 1996).

[119] For more details on the methodology, see the *Renewable Fuels Module of the National Energy Modeling System, Model Documentation 2000*, DOE/EIA-M069(2000), January 2000.

[120] Biocycle, *The State of Garbage in America*, Annual Series (April 1988 - April 1997).

Notes and Sources

[121] 1996 estimate from the U.S. Environmental Protection Agency (EPA), *Characterization of Municipal Solid Waste in the United States: 1997 Update*, Report#: EPA530-R-98-007(Prairie Village, KS, May 1998) (Table 34).

[122] U.S. Department of Energy, *The Climate Change Action Plan: Technical Supplement*, DOE/PO-0011, (Washington, DC, March 1994) p. 57.